# Association of Reptilian and Amphibian Veterinarians Conference (ARAV 2019)

Held at ExoticsCon 2019

St. Louis, Missouri, USA
29 September – 3 October 2019

ISBN: 978-1-7138-0337-9

**Printed from e-media with permission by:**

Curran Associates, Inc.
57 Morehouse Lane
Red Hook, NY 12571

Some format issues inherent in the e-media version may also appear in this print version.



Printed with permission by Curran Associates, Inc. (2020)

For permission requests, please contact Association of Reptilian & Amphibian Veterinarians at the address below.

Association of Reptilian & Amphibian Veterinarians
5003 SW 41st Blvd
Gainesville FL 32608-4930
USA

Phone: 480-703-4941

info@arav.org

**Additional copies of this publication are available from:**

Curran Associates, Inc.
57 Morehouse Lane
Red Hook, NY 12571 USA
Phone: 845-758-0400
Fax: 845-758-2633
Email: curran@proceedings.com
Web: www.proceedings.com

# TABLE OF CONTENTS

# EXOTICSCON 2019
# PRELIMINARY PROGRAM GUIDE

St. Louis, MO • Hyatt Regency St. Louis at the Arch • September 29 - October 3, 2019

BROUGHT TO YOU BY

**EXOTICSCON.ORG**
BUILDING EXOTICS EXCELLENCE.
ONE CITY. ONE CONFERENCE.

Association of Avian Veterinarians

AEMV
ASSOCIATION OF EXOTIC MAMMAL VETERINARIANS

ARAV
ASSOCIATION OF REPTILIAN AND AMPHIBIAN VETERINARIANS

# SUNDAY, SEPTEMBER 29

*Please note, this grid is subject to change slightly as speaker conflicts arise.*
*A copy of the completed grid will be provided at time of registration on-site.*

| HANDS-ON WORKSHOPS (HOW) | AAV WORKSHOP | AAV WORKSHOP | AAV WORKSHOP (AM) ARAV WORKSHOP (PM) | AAV WORKSHOP | ARAV WORKSHOP (AM) AAV WORKSHOP (PM) | ARAV WORKSHOP |
|---|---|---|---|---|---|---|
| LOCATION | GRAND H | MILLS 6 | GRAND G | STERLING 2 | STERLING 1 | GRAND F |
| SPONSOR | | | | | | |
| 7:00-8:00 AM | **Breakfast On Your Own** | | | | | |
| 8:00-12:00 AM | **2923** Surgery of the Respiratory System, Gastrointestinal System & Reproductive Tract *Brian Speer* | **2823** Avian Handling Techniques: Creating the Low Stress Avian Vet *Melody Hennigh* | **2620** Avian Endoscopy & Visceral Biopsy: The Key to a Definitive Diagnosis! *Stephen Divers & Scott Stahl* | **2922** Tips & Tricks in Avian Medicine *Anna Osokfsy & Jennifer Graham* | **2940** Reptile Ultrasound *Shannon Ferrell* | **2811** Shell Repair for Aquatic Freshwater Turtles: Beyond Resin & Fiberglass! *Renee Schott* |
| 12:00-1:30 PM | **Lunch On Your Own** | | | | | |
| 1:30-5:30 PM | (Continues) **2923** Surgery of the Respiratory System, Gastrointestinal System & Reproductive Tract *Brian Speer* | (Repeat) **2823R** Avian Handling Techniques: Creating the Low Stress Avian Vet *Melody Hennigh* | **2619** Lizard Endoscopy/ Endosurgery *Stephen Divers & Scott Stahl* | | **2935** Avian Ultrasound *Maria Evola* | **2909** Prefemoral Reproductive Surgery in Female Aquatic Chelonians *Nicola Di Girolamo* |
| 4:30-5:00 PM | **AAV JAMS Committee, Location: Mills 2** | | | | | |
| 6:30-9:30 PM | **Welcome Reception, Location: City Museum** | | | | | |

**REGISTRATION – REGENCY COAT ROOM – OPENS 7:30 AM**

## Sunday, September 29 (cont.)

| HANDS-ON WORKSHOPS (HOW) | ARAV WORKSHOP | ARAV WORKSHOP | AEMV WORKSHOP | AEMV WORKSHOP | AEMV WORKSHOP | AEMV WORKSHOP |
|---|---|---|---|---|---|---|
| LOCATION | MILLS 7 | STRELING 6 | STERLING 7 | MILLS 1 | MILLS 3 | GRAND C |
| 7:00-8:00 AM | **Breakfast On Your Own** | | | | | |
| 8:00-12:00 AM | **2796**<br>Reptile & Amphibian Nutrition & Nutritional Diseases<br>*Thomas Boyer* | **2710**<br>Current Amphibian Clinical Medicine & Diagnostics<br>*Colin McDermott & Daniel Dombrowski* | **2731**<br>Chiropractic Care & Acupuncture in Rabbits<br>*Robert Ness* | **2636**<br>How to Intubate Almost Any Exotic Companion Mammal<br>*Stacey Leonatti-Wilkinson & Salina Locke* | **2726**<br>Elective Altering: Options & Alternatives for Exotic Companion Mammals<br>*Angela Lennox & Jean-Francois Quniton* | **2758**<br>Ear Surgeries in Rabbits & Rodents<br>*Charly Pignon* |
| 12:00-1:30 PM | **Lunch On Your Own** | | | | | |
| 1:30-5:30 PM | **2667**<br>Recycling for Reptiles: Utilizing Commonly Available Materials to Enrich Your Captive Reptile's Life<br>*Edythe Sontag* | **2656**<br>Introduction to Sea Turtle Medicine<br>*Charles Innis* | **2942**<br>Exotic Animal Hematology<br>*Susan Fielder & Ian Kanda* | (Repeat)<br>**2636R**<br>How to Intubate Almost Any Exotic Companion Mammal<br>*Stacey Leonatti-Wilkinson & Salina Locke* | (Repeat)<br>**2726R**<br>Elective Altering: Options & Alternatives for Exotic Companion Mammals<br>*Angela Lennox & Jean-Francois Quniton* | (Repeat)<br>**2758R**<br>Ear Surgeries in Rabbits & Rodents<br>*Charly Pignon* |
| 4:30-5:00 PM | **AAV JAMS Committee, Location: Mills 2** | | | | | |
| 6:30-9:30 PM | **Welcome Reception, Location: City Museum** | | | | | |

REGISTRATION – REGENCY COAT ROOM - OPENS 7:30 AM

# MONDAY, SEPTEMBER 30

*Please note, this grid is subject to change slightly as speaker conflicts arise.*
*A copy of the completed grid will be provided at time of registration on-site.*

REGISTRATION – REGENCY COAT ROOM – OPENS 7:30 AM

| | **Plenary Session: Sponsored by Lafeber Emeraid** |
|---|---|
| LOCATION | REGENCY ABCDEF |
| 7:00-8:00 AM | **Continental Breakfast with Posters, Location: Grand CDEF Pre-Function** |
| 8:00-8:15 AM | **Plenary Session - Welcome**<br>*Nicola di Girolamo, Dr. med. vet., GPCert (ExAP), MSc (EBHC), PhD, DECZM (Herpetology); Tim Tristan, DVM, DABVP (Reptile & Amphibian); Yvonne van Zeeland, DVM, MVR, PhD, DECZM (Avian & Small Mammal), CPBC* |
| 8:15-8:35 AM | **2760** Plenary 1-AAV: Gonadal Activity Suppression in Chickens, Rats, & Swine Using an Antibody-Guided Lipid-Based Nanocomplex<br>*Jennifer Graham* |
| 8:35-8:55 AM | **2766** Plenary 2-AEMV: Evaluation of the Effects of Metronidazole on Food Intake in Chinchillas<br>*Dustin Fink* *2018 Reserach Grant Award Receipient* |
| 8:55-9:05 AM | **AEMV, Oxbow Animal Health: Exotic Mammal Health Award** |
| 9:05-9:25 AM | **2695** Plenary 3-ARAV: Erythrocyte Sedimentation Rate & Hemoglobin-Binding Protein in Eastern & Ornate Box Turtles (*Terrapene spp.*)<br>*Laura Adamovicz* |
| 9:25-9:45 AM | **2631** Plenary 4-AAV: Avian Sources for Transmission of Psittacosis to Humans<br>*Yvonne van Zeeland* |
| 9:45-10:00 AM | **AAV, LaFeber Avian Practitioner of the Year Award** |
| 10:10-10:30 AM | **Coffee Break** |
| 10:30-10:50 AM | **2682** Plenary 5-AEMV: Efficacy & Safety of Buprenorphine in African Pygmy Hedgehogs (*Atelerix albiventris*)<br>*Grayson Doss* *2017 Reserach Grant Award Receipient* |
| 10:50-11:10 AM | **2806** Plenary 6-ARAV: Left-Shifting in Non-Mammalian Vertebrate Heterophils & Neutrophils: Quantification & Clinical Significance<br>*Nicole Stacy* |
| 11:10-11:20 AM | **ARAV, Zoo Med Laboratories Reptile & Amphibian Veterinary Excellence Award** |
| 11:20-12:00 PM | **Keynote Address:** The Saint Louis Zoo's WildCare Institute - A Zoo-Based Conservation Program<br>*R. Eric Miller, DVM, DACZM, Executive Director, WildCare Institute, Saint Louis Zoo* |
| 12:00-1:30 PM | **AAV, AEMV, ARAV Joint Business Lunch (Lunch Included with Registration), Location: Regency ABCDEF** |

REGISTRATION – REGENCY COAT ROOM

| | **AAV Track** | **Hybrid Track** | **ARAV Track** | **Technician Track** | **AEMV Track** |
|---|---|---|---|---|---|
| LOCATION | REGENCY C | REGENCY D | REGENCY E | REGENCY F | REGENCY A & B |
| SPONSOR | Lafeber's EmerAid | | UMIAMI Avian & Wildlife Laboratory | | |
| | **Cutting Edge Medicine** | | **Infectious Disease & Clinical Procedures** | | **Clinical Pathology** |
| 2:15-2:30 PM | **2677**<br>Use of a Vessel Sealing Device for Ovariectomy in Birds - Chicken Model<br>*Avery Bennett* | **2740**<br>Common Referral Issues for Reptiles, Small Mammals & Birds: Cases to Learn From<br>*Teresa Bradley Bays* | **2839**<br>Prevalence of Herpesviruses in Chelonians in Veterinary Practice in Europe<br>*Rachel Marschang* | **2680**<br>Decreasing Patient Stress Before, During & After the Veterinary Visit<br>*Sarah Dehn* | **2651**<br>Plasma Refractometric TS & TP by the Biuret Assay Comparison in Pet Rabbits & Ferrets<br>*David Eshar* |
| 2:30-2:35 PM | **Break to Move to Next Session** | | **Break to Move to Next Session** | | **Break to Move to Next Session** |
| 2:35-2:50 PM | **2924**<br>Comparison of Different Doses & Delivery Methods of IsoVue 370 IV for CT Contrast<br>*Nick Kirk* | | **2881**<br>A Novel *Actinomyces* Associated with Granulomatous Cloacal & Hemipenal Inflammation in Male Ball Pythons<br>*Robert Ossiboff* | | **2918**<br>Types of Ferret Uroliths Submitted to Four Laboratories from 2010-2018<br>*Eryn Hanak* |
| 2:50-2:55 PM | **Break to Move to Next Session** | | **Break to Move to Next Session** | | **Break to Move to Next Session** |
| 2:55-3:10 PM | **2756**<br>Diagnosis & Treatment of *Gordonia* Species Infection in a Lovebird (*Agapornis*)<br>*Nicolle Lofgren* | | **2903**<br>Detection of New Nidoviruses & an Orthoreovirus in Association with Respiratory Disease in Veiled Chameleons<br>*Anke Stoehr* | | **2660**<br>Defining Anitgens of Interest in the Serological Reponse to *E. cuniculi* in Rabbits<br>*Carolyn Cray* |
| 3:00-3:30 PM | **Coffee Break**<br>*(ExoticsCon Attendees will have 10 minutes for this break between sessions)* | | | | |

## Monday, September 30 (cont.)

| | AAV Track | Hybrid Track | ARAV Track | Technician Track | AEMV Track |
|---|---|---|---|---|---|
| LOCATION | REGENCY C | REGENCY D | REGENCY E | REGENCY F | REGENCY A & B |
| SPONSOR | Lafeber's EmerAid | | UMIAMI Avian & Wildlife Laboratory | | |
| | **Cutting Edge Medicine** | | **Infectious Disease & Clinical Procedures** | | **Research** |
| 3:10-3:20 PM | **Break to Move to Next Session** | | | | |
| 3:20-3:35 PM | **2637** Evaluation of a Pressure Sensitive Walkway for Objective Gait Analysis in Normal & Arthritic Ducks *Julie Sheldon* | **2718** 10 Things You Didn't Know About a More Efficient Use of Google Scholar for Your Research *Christian Gumpenberger* | **2916** Effects of Fecal Sample Age On Direct Microscopic Examination of Protozoan Parasites *Anna-Marie Ruoff* | **2696** Maximizing the Healing Power of Exotic Animal Nursing *Jovanna Berls* | **2742** Copper Concentration in 10 Popular Commercial Ferret Diets *Miranda Torkelson* |
| 3:35-3:40 PM | **Break to Move to Next Session** | | | | **Break to Move to Next Session** |
| 3:40-3:55 PM | **2856** Regional Limb Perfusion in a Pekin Duck (*Anas platyrhynchos*) with Severe Unilateral Pododermatitis *Meera Music* | **2857** Stayin' Alive: Current Topics in Exotic Emergency Medicine *Alicia McLaughlin* | **2838** The Subcarapacial Vessel in Chelonians: Safe for Intravenous Injections? *Kelly Rockwell* | | **2761** Factors Associated with Longevtity in Pet Rabbits *Emma Keeble* |
| 3:55-4:00 PM | **Break to Move to Next Session** | | **Break to Move to Next Session** | | **Break to Move to Next Session** |
| 4:00-4:15 PM | **2851** Adverse Effects of Multi-Dose Oral Meloxicam Administration in Rhode Island Red Chickens *Emma Houck* | | **2917** Indications, Outcomes & Complications of Prefemoral Reproductive Surgery in Client-Owned Chelonians: -22 Cases, 2017-2018 *Nicola Di Girolamo* | | **2659** Comparison of Axillary & Inguinal Temperature to Rectal Temperature in Health Guinea Pigs (*Cavia porcellus*) *Ivana Levy* |
| 4:15-4:20 PM | **Break to Move to Next Session** | | **Break to Move to Next Session** | | |
| 4:20-4:35 PM | **2834** Pharmacokinetics of Oral & Intravenous Administration of Trimethoprim-Sulfamethoxazole in Rhode Island Red Chickens *Olivia Petritz* | | **2866** Effect of Lithium Heparin Versus EDTA on the Hematology of Ball Pythons *Lauren Schmidt* | **2836** Trouble Shooting the Slow Recovery *Sarah Dehn* | **2911** 20-Year Trend in Type of References for Small Mammals Drug Dosages in Exotic Formulary *Stephanie Kline* |
| 4:35-4:40 PM | **Break to Move to Next Session** | | **Break to Move to Next Session** | | **Break to Move to Next Session** |
| 4:40-4:55 PM | **2618** Pharmacokinetics & Drug Residues in Eggs Following Clavamox® Oral Administration to Domestic Chickens *Marcy Souza* | | **2914** 20-Year Trend in Type of References Used for Reptile Drug Dosages in Exotic Animal Formulary *Marion Carron* | | **2749** Study of Efficacy of Rabies Vaccine in Rabbits: Immunity & Safety *Chaowaphan Yinharnmingmongkol* |
| 4:55-5:00 PM | **Break to Move to Next Session** | | **Break to Move to Next Session** | | **Break to Move to Next Session** |
| 5:00-5:15 PM | **2762** Amlodipine for Treatment of Hypertension in Psittacines: Six Cases (2010-2018) *Dustin Fink* | | **2919** 20-Year Trend in Type of References Used for Reptile Reference Intervals in Exotic Animal Formulary *Cassandra Smead* | | **2728** Non-Invasive Blood Pressure Measurement in Rabbits Using High Definition Oscillometry *Nico Schoemaker* |
| 5:30-7:30 PM | **Student Reception, Location: Grand A & D** | | | | |
| 7:00-9:00 PM | **Oxbow Animal Health Reception (Invitation Only), Location: Gateway East** | | | | |
| LOCATION | AAV Committees - REGENCY C | AAV Committees - REGENCY C | ARAV Committees - REGENCY A | ARAV Committees - REGENCY A | AEMV Committees - REGENCY B |
| 7:30-8:30 PM | Conservation Committee | Education Committee | Infectious Disease Committee | Memberships & PR Committee | Finance Committee |
| 7:30-8:30 PM | International Committee | Membership Committee | Green Conference Committee | Education Committee | Marketing Committee |
| 7:30-8:30 PM | Research Committee | Welfare Committee | | | |
| 8:30-9:30 PM | Aviculture Committee | Finance & Development Committee | International Committee | LAW Committee | International, Student & Membership Committee |
| 8:30-9:30 PM | Legislative Committee | Marketing & Social Media Committee | Historic Committee | Technician Committee | Research Committee |
| 8:30-9:30 PM | Students & Vet Schools Committee | Technician Committee | | | |

REGISTRATION – REGENCY COAT ROOM

# TUESDAY, OCTOBER 1

*Please note, this grid is subject to change slightly as speaker conflicts arise.*
*A copy of the completed grid will be provided at time of registration on-site.*

REGISTRATION – REGENCY COAT ROOM - OPENS 7:30 AM

EXHIBIT HALL - MOVE-IN, REGENCY BALLROOM - 9:00 AM - 5:00 PM

| | AAV Track 1 | AAV Track 2 | ARAV Track | Hybrid Track | Technician Track | AEMV Track |
|---|---|---|---|---|---|---|
| LOCATION | REGENCY C | GRAND A | GRAND D | GRAND C | GRAND B | REGENCY A & B |
| SPONSOR | | | Lafeber's EmerAid | | | Lafeber's EmerAid |
| 7:00-8:00 AM | **Continental Breakfast with Posters, Location: Grand CDEF Pre-Function** | | | | | |
| | **AEMV & ARAV Past President's Breakfast** Location: Gateway East | | | | | |
| | **AAV Past President's Breakfast** Location: Gateway West | | | | | |
| | **Flock Health Management** | **Wild Bird Health** | **Case Reports** | | | **Respiratory & Gastroenterology** |
| 8:00-8:15 AM | **2860** Cassowary Eggs: Collection, Incubation & Hatchability *April Romagnano* | **2768** Use of Intravenous Lipid Emulsion Therapy as a Novel Treatment for Brevetoxicosis in Double-Crested Cormorants *Heather Barron* | **2893** Effectiveness of Common Disinfecting Agents Against Isolates of *Nannizziopsis guarroi* *Caitlyn Hemby* | **2892** Interstate Movement of Exotic Pets & the Role of Veterinarians in Managing Risk *Jennifer Ballard* | **2927** Introduction to Foraging *Melody Hennigh* | **2765** Acute Lung Injury (ALI) & Acture Respiratory Distress Syndrome (ARDS) in Two Rabbits (*Oryctolagus cuniculus*) *Julianne Richard* |
| 8:15-8:20 AM | **Break to Move to Next Session** | | | | | **Break to Move to Next Session** |
| 8:20-8:35 AM | **2859** Four Infertile Female Cassowaries Rendered Fertile After Dramatic Diet Change to a Highly Frugivore-Based Diet *April Romagnano* | **2815** Lessons Learned from a Fuel Spill in Frigid Northern Weather *Renee Schott* | **2621** Treatment of a Retrobulbar Adenocarcinoma Using Surgery & Radiation Therapy in a *Morelia spilota* *Joshua Llinas* | **2751** To See or Not to See: Diagnostic Imaging in Giant African Land Snails *Michaela Gumpenberger* | | **2727** Diagnosis & Treatment of a Porto-Systemic Shunt in a Ferret *Nico Schoemaker* |
| 8:35-8:40 AM | **Break to Move to Next Session** | | | | | **Break to Move to Next Session** |
| 8:40-8:55 AM | **2730** Investigation of Ionized Calcium & Vitamin D3 in African Penguins *Jennilee Morrison* | **2675** Pre-Release Hematologic Values in Rehabilitated Mississippi Kites (*Ictinia mississippiensis*) *Ian Kanda* | **2795** Septic Arthritis in a Bearded Dragon Caused by *Mycobacterium* *Julian Vaillard* | **2929** Advanced Diagnostic Imaging *Scott Echols* | **2928** Parrots & Children: Expanding the Flock *Melody Hennigh* | **2910** Caudoventral Abdominal Wall Herniation in Male Domestic Rabbits (*Oryctolagus cuniculus*) *Stephanie Kline* |
| 8:55-9:00 AM | **Break to Move to Next Session** | | | | | **Break to Move to Next Session** |
| 9:00-9:15 AM | **2679** Utilizing Dietary Nutritional Analysis in the Investigation of Developmental Skeletal Abnormalities in Captive Peregrine Falcons *Thomas Dutton* | **2782** Investigation of the Avian Cornea Using Anterior Segment Optical Coherence Tomography (AS-OCT) & Confocal Microscopy *Ruediger Korbel* | **2798** Successful Management of Anemia & Hemocoelom in a Bearded Dragon Using Multiple Homologous Whole-Blood Transfusions *Laura Adelina Aguilar* | | **2863** Low-Stress Handling Techniques of Small Mammals with a Focus on Rabbits & Guinea Pigs *Stephanie Kotschwar* | **2861** Successful Decompression Technique in Rabbits Presenting with Gastric Outflow Obstructions *Bill Guerrera* |
| 9:15-9:20 AM | **Break to Move to Next Session** | | | | | **Break to Move to Next Session** |
| 9:20-9:35 AM | **2627** Post-Mortem Findings in a Collection of Captive Raptors *John Chitty* | **2925** Padre Peregrine Falcon Survey on South Padre Island *Michael Jones* | **2829** Adenocarcinoma of the Membrana Nictitans in an Inland Bearded Dragon (*Pogona vitticeps*) *Lauren Schmidt* | | | **2643** Cholangioma in a Degu (*Octodon degus*) *Monica Sidorowicz* |
| 9:35-9:50 PM | **Coffee Break** | | | | | |

## Tuesday, October 1 (cont.)

REGISTRATION - REGENCY COAT ROOM - OPENS 7:30 AM

EXHIBIT HALL - MOVE-IN, REGENCY BALLROOM - 9:00 AM - 5:00 PM

| | AAV Track 1 | AAV Track 2 | ARAV Track | Hybrid Track | Technician Track | AEMV Track |
|---|---|---|---|---|---|---|
| LOCATION | REGENCY C | GRAND A | GRAND D | GRAND C | GRAND B | REGENCY A & B |
| | | | **Case Reports** | | | **Oncology** |
| SPONSOR | | | Lafeber's EmerAid | | | Lafeber's EmerAid |
| 9:50-10:05 AM | **2713**<br>Vertical Transmission of Avypolyomavirus<br>*Glenda Murciano* | **2827**<br>Optimizing Outcomes in Wildlife Rehabilitation - A Comprehensive Approach<br>*Michelle Willette* | **2792**<br>Diagnosis & Treatment of Proliferative Fungal (*Fusarium proliferatum*) Stomatitis in a Bearded Dragon (*Pogona vitticeps*)<br>*Rosa Bañuelos* | **2757**<br>Tips & Tricks to Start Acupuncture in Exotic Species<br>*Warinda Somrith* | **2719**<br>Avian Emergency Triage & the Role of the Veterinary Technician<br>*Liz Marshall* | **2652**<br>Treatment of Hyperthyroidism with Radiation Therapy (I131) in Three Guinea Pigs (*Cavia porcellus*)<br>*Brynn McCleery* |
| 10:05-10:10 AM | **Break to Move to Next Session** | | **Break to Move to Next Session** | | | |
| 10:10-10:25 AM | **2624**<br>Prevalence, Sample & Test Performance for *Chlamydia Psittaci* in Psittacine Birds<br>*Yvonne van Zeeland* | | **2722**<br>Don't Poke the Dragon: Tail Base Aneurysm in an Inland Bearded Dragon (*Pogona vitticeps*)<br>*Nicole Furst* | **2902**<br>Applying Local Blocks Across the Taxons<br>*Angela Lennox* | **2921**<br>Using Natural History Information to Raise the Bar of Reptile Husbandry<br>*Ian Kanda* | **2879**<br>Sudden Death During ACTH Stimulation Test in a Clinically Healthy Pet Rabbit<br>*Nicola Di Girolamo* |
| 10:25-10:30 AM | **Break to Move to Next Session** | | | | | **Break to Move to Next Session** |
| 10:30-10:45 AM | **2672**<br>Potential Risks of Fungal Respiratory Disease in Companion Avian Species From Window-Mounted Air Conditioners<br>*Laura Wade* | **2684**<br>Myelolipomatosis in Javan Pond Herons<br>*Yami Lago-Alvarez* | **2753**<br>Surgical Correction of Urinary Bladder Prolapses in Burmese Star Tortoise<br>*Chaowaphan Yinharnmingmongkol* | | | **2736**<br>Use of Raditaion Therapy in African Pygmy Hedgehog with Liposarcoma<br>*Polina Maratkanova* |
| 10:45-10:50 AM | **Break to Move to Next Session** | | | | | **Break to Move to Next Session** |
| 10:50-11:05 AM | **2720**<br>Using Metagenomic Sequencing to Identify a Novel Viral Agent Associated with Disease in a Psittacine<br>*Susan Clubb* | **2606**<br>Discospondylitis in a Greater Flamingo (*Phoenicopterus roseus*)<br>*Shangzhe Xie* | **2752**<br>Urinary Retention Demonstrated by Means of Diagnostic Imaging in a Caiman Lizard (*Dracaena guianensis*)<br>*Michaela Gumpenberger* | | | **2642**<br>Squamous Cell Carcinoma of Cheek Muscles in a Degu (*Octodon degus*)<br>*Monica Sidorowicz* |
| 11:05-11:10 AM | **Break to Move to Next Session** | | | | | |
| 11:10-11:25 AM | **2623**<br>Nature Calls – Ecological Determinants for Abnormal Behaviour & Breeding Problems in Captive Psittacines<br>*Yvonne van Zeeland* | **2799**<br>Success of Avian Orthopedic Management Techniques<br>*Dana Franzen-Klein* | **2769**<br>Necrotizing Fasciitis & Osteomyelitis Due to *Fusarium Solani* in a Russian Tortoise (*Agrionemys horsfieldii*)<br>*Jana Mazor-Thomas* | **2807**<br>Anesthesia of Small Urban Wildlife<br>*Renee Schott* | **2635**<br>Introduction to Basic Raptor Techniques<br>*Marcia Gannon* | **2858**<br>Review of Squamous Cell Carcinoma in the Domestic Rabbit (*Oryctolagus cuniculus*)<br>*Alicia McLaughlin* |
| 11:25-11:30 AM | **Break to Move to Next Session** | | | | | **Break to Move to Next Session** |
| 11:30-11:45 AM | **2788**<br>Identifying Characteristics to Aid Social Pairing of Captive Parrots<br>*Jamie Douglas* | **2926**<br>Outcome of Antebrachial Fractures in Birds of Prey: 620 Cases (1986-2015)<br>*Claire Vergneau-Grosset* | **2785**<br>Successful Diagnosis & Management of a Plastron Eosinophilic Pseudocyst with a Partial Plastronectomy<br>*Sean Perry* | | | **2773**<br>Intrascrotal Anaplastic Sarcoma & Bilateral Testicular Necrosis in a Guinea Pig (*Cavia porcellus*)<br>*Jana Mazor-Thomas* |
| 11:45-11:50 AM | **Break to Move to Next Session** | | | | | **Break to Move to Next Session** |
| 11:50-12:05 AM | **2775**<br>The Pellet Puzzle: An Evaluation of Three Proposed Pellet Conversion Methods<br>*Alison Cummings* | **2697**<br>Effects of Perching Surfaces & Foot Bandaging on the Metatarsal Pad in the Peregrine Falcon<br>*Trinita Barboza* | **2855**<br>Management of Respiratory Infection in a Reticulated Python (*Python reticulatus*) Using an Osmotic Antibiotic Pump<br>*Meera Music* | | | **2683**<br>Clinico-Pathologic Features of Vascular Neoplasia in Hedgehogs<br>*Michael Garner* |
| 12:05 PM | **Transportation to the St. Louis Zoo - Lunch at the Zoo until 1:30 PM** | | | | | |
| 7:00-9:00 PM | **Dinner, Live & Silent Auctions at the Zoo** | | | | | |

# WEDNESDAY, OCTOBER 2

*Please note, this grid is subject to change slightly as speaker conflicts arise.*
*A copy of the completed grid will be provided at time of registration on-site.*

| | AAV Track 1 | AAV Track 2 | AAV Track 3 | AAV Track 4 |
|---|---|---|---|---|
| LOCATION | REGENCY A | REGENCY B | REGENCY C | GRAND B |
| SPONSOR | | Lafeber's EmerAid | | |
| 7:00-8:00 AM | **Continental Breakfast for All, Location: Exhibit Hall** | | | |
| 7:00-8:00 AM | **AAV Board of Directors Breakfast (Invitation Only), Location: Sterling 5** | | | |
| | **Wild Bird Health Management** | **Internal Medicine** | **Aviculture & Flock Management** | **All Things Visible** |
| 8:00-9:00 AM<br>9:00-9:05 AM<br>9:05-10:05 AM | **2778**<br>Avian Wound Management<br>*Heather Barron & Rebecca Duerr* | **2772**<br>The Renal & Osmoregulatory Systems of Birds: Anatomy & Physiology<br>*Susan Orosz & Scott Echols* | **2717**<br>Fundamentals of Aviculture for Veterinarians<br>*Susan Clubb* | **2617**<br>Introduction to Diagnostic Imaging of Raptors & Companion Birds<br>*Seth Oster* |
| 10:05-10:45 AM | **Exhibit Hall Break** | | | |
| 10:45-11:45 AM | **2801**<br>Avian Physiotherapy & Reconditioning in a Rehabilitation Program<br>*Dana Franzen-Klein* | **2704**<br>Avian Oncology, Current Diagnostic & Treatment Options<br>*Thomas Tully* | **2604**<br>A Bird in Hand is Worth Two in the Bush: Flock Management of Disease Outbreaks<br>*Shangzhe Xie* | **2632**<br>Studies on Parrot Intelligence & Cognitive Abilities & Their Effect on Daily Practice<br>*Irene Pepperberg* |
| 11:45 AM-1:00 PM | **Lunch On Own** | | | |
| 1:00-2:00 PM<br>2:00-2:05 PM<br>2:05-3:05 PM | **2764**<br>Floods, Fires, Spills & Blooms- The Avian Practitioner & Natural Disasters<br>*Michelle Hawkins & Heather Barron* | **2653**<br>Practical Overview of Poultry Diseases You Need to Know<br>*Cheryl Greenacre* | **2877**<br>The Incredible Egg: Physiology & Diagnostics<br>*J. Jill Heatley* | **2777**<br>Avian Ophthalmology: Principles & Application<br>*Ruediger Korbel* |
| 3:05-3:30 PM | **Exhibit Hall Break** | | | |
| 3:30-4:30 PM | **2616**<br>Introduction to Hematology of Raptors & Companion Birds<br>*Seth Oster* | **2671**<br>Managing the Avian Microbiome in Health & Disease<br>*Bob Dahlhausen* | **2847**<br>Psittacine Neonatology & Pediatrics: From the Egg to the Flock<br>*Mikel Sabater* | **2685**<br>Gross Lesion Recognition for the Avian Clinician<br>*Michael Garner* |
| 4:30-4:35 PM | **Break to Move to Next Session** | | | |
| 4:35-5:35 PM | **2812**<br>Wildlife Rehabilitation: Common Songbird Injuries, Illness & Treatment<br>*Renee Schott* | **2803**<br>Workshop: Communication with Birds Using Phonis & Consent Training<br>*Jennifer Cunha & Susan Clubb* | **2626**<br>Management, Medicine & Surgery of Captive Penguins<br>*John Chitty* | |
| 3:30-5:00 PM | **AAV Conservation Forum, Location: Sterling 5** | | | |
| 5:30-6:30 PM | **ABVP Reception, Location: Mills 3** (CAITEC) | | | |
| 7:00-9:00 PM | **Trivia Challenge, Sponsored by Zupreem, Location: Gateway West** (ZuPreem) | | | |

REGISTRATION – REGENCY COAT ROOM – OPENS 7:30 AM

EXHIBIT HALL OPEN - REGENCY BALLROOM - 9:00 AM - 5:00 PM

## Wednesday, October 2 (cont.)

REGISTRATION – REGENCY COAT ROOM – OPENS 7:30 AM

EXHIBIT HALL OPEN - REGENCY BALLROOM - 9:00 AM - 5:00 PM

| | ARAV Track | AEMV Track 1 | AEMV Track 2 |
|---|---|---|---|
| LOCATION | GRAND C | GATEWAY EAST | GATEWAY WEST |
| SPONSOR | | | |
| 7:00-8:00 AM | **Continental Breakfast for All, Location: Exhibit Hall** | | |
| 7:00-8:00 AM | **AAV Board of Directors Breakfast (Invitation Only), Location: Sterling 5** | | |
| 8:00-9:00 AM | **2875** Techniques for Minimally-Invasive Removal of Urinary Stones in Chelonians *Nicola Di Girolamo* | **2848** Small Mammal Ophthalmology: Anatomy, Physiology, Diagnostics & Therapeutics *Mikel Sabater* | **2733** Practical Chiropractic & Acupuncture Techniques in Rabbits for Clinicians *Robert Ness* |
| 9:00-9:05 AM | **Break to Move to Next Session** | | |
| 9:05-10:05 AM | **2641** Urolithiasis & Surgery of the Lower Urinary Tract in Reptiles *Heather Bjornebo* | **2706** Loco-Regional Anaesthesia in Rabbits *Stefania Scarabelli* | **2784** Husbandry, Medicine & Surgery of Pet Mini Pigs *Louisa Asseo* |
| 10:05-10:45 AM | **Exhibit Hall Break** | | |
| 10:45-11:45 AM | **2687** Gross Lesion Recognition for the Reptile & Amphibian Practitioner *Michael Garner* | **2837** Not Just Neoplasia: The Diagnosis & Treatment of Reproductive Diseases in Exotic Companion Mammals *Tracey Ritzman* | **2912** Tips to Write Articles for the *Journal of Exotic Pet Medicine* *Nicola Di Girolamo* |
| 11:45 AM-1:00 PM | **Lunch On Your Own** | | |
| 1:00-2:00 PM | **2705** Chelonian Total Intravenous Anaesthesia in Clinical Practice *Stefania Scarabelli* | **2897** Review of Urinary Tract Disease of Selected Exotic Mammals *Angela Lennox* | **2686** Gross Lesion Recognition for the Small Mammal Clinician *Michael Garner* |
| 2:00-2:05 PM | **Break to Move to Next Session** | | |
| 2:05-3:05 PM | **2849** Reptile Ophthalmology: Anatomy, Physiology, Diagnostics & Therapeutics *Mikel Sabater* | **2701** Hand-Rearing Small Mammals: Rabbits, Rats, Guinea Pigs & Ferrets *Anne Fowler* | |
| 3:05-3:30 PM | **Exhibit Hall Break** | | |
| 3:30-4:30 PM | **2880** Ophidian Nidoviruses: What We Know, What We Don't & What We are Figuring Out *Robert Ossiboff* | **2894** Options for Surgical Management of Intestinal Foreign Bodies in the Rabbit *Angela Lennox* | **2673** African Pygmy Hedgehogs: Husbandry & Clinical Management *Grayson Doss* |
| 4:30-4:35 PM | **Break to Move to Next Session** | | |
| 4:35-5:35 PM | **2633** Reptile Emergency & Critical Care: Current Updates from the Literature *Olivia Petritz* | **2832** Comprehensive Review of Ferret Gastrointestinal Diseases *Ivy Thomas* | |
| 3:30-5:00 PM | **AAV Conservation Forum, Location: Sterling 5** | | |
| 5:30-6:30 PM | **ABVP Reception, Location: Mills 3** | | |
| 7:00-9:00 PM | **Trivia Challenge, Sponsored by Zupreem, Location: Gateway West** | | |

# THURSDAY, OCTOBER 3

*Please note, this grid is subject to change slightly as speaker conflicts arise.*
*A copy of the completed grid will be provided at time of registration on-site.*

REGISTRATION – REGENCY COAT ROOM – OPENS 7:30 AM

EXHIBIT HALL – REGENCY BALLROOM – 9:00 AM - 5:00 PM

| | AAV Track 1 | AAV Track 2 | ARAV Track 1 | ARAV Track 2 | AEMV Track |
|---|---|---|---|---|---|
| LOCATION | PARK VIEW | GRAND B | GRAND C | REGENCY C | REGENCY A & B |
| SPONSOR | | | | | |
| 7:00-8:00 AM | **Continental Breakfast for All, Location: Exhibit Hall** | | | | |
| 8:00 AM-12:00 PM | **Nature Outing, Location: Cliff Cave Park** *(Additional fee. Limited attendance. Departs hotel at 7:30 AM and returns by 12:00 PM for attendees to eat prior to afternoon sessions.)* | | | | |
| | **Oncology** | **Pharmacology & Anatomy** | **Specialist Medicine** | **Conservation & Wildlife** | **Surgery** |
| 8:00-8:15 AM | **2711** Carcinoma in a Green-Winged Macaw (*Ara chloropterus*) *Peggy Shashy* | **2930** Decision Making in Avian Therapeutics – Pharmacological Principles Underlying Drug & Drug Dose Selection *Keven Flammer* | **2744** Reptilian & Amphibian Reproductive Cycles, Eggs & Embryos, Normal Development & Incubation *Jeanette Wyneken* | **2725** Diet as a Cause of Gastroduodenal Perforations in Captive Spiny Softshell Turtles (*Apalone spinifera*) *Shannon Ferrell* | **2622** Minimally Invasive Ovariectomy in Guinea Pigs: A Pilot Study Using Mini-Laparoscopic Instrumentation *Julianne McCready* |
| 8:15-8:20 AM | **Break to Move to Next Session** | | | **Break to Move to Next Session** | |
| 8:20-8:35 AM | **2907** Diagnosis & Management of Arrhythmia Associated with a Heart-Based Tumor in a Yellow-Headed Amazon Parrot *Abigail Duvall* | | | **2654** Effect of Bone Morphogenetic Protein on Chelonian Experimental Carapace Defects *Charles Innis* | **2876** Appendectomy in a Rabbit with Chronic Appendix Impaction & Necrosis *Nicola Di Girolamo* |
| 8:35-8:40 AM | **Break to Move to Next Session** | | | **Break to Move to Next Session** | |
| 8:40-8:55 AM | **2895** Radiation Therapy for Treatment of Thymoma in an African Grey Parrot *Angela Lennox* | **2692** Pharmacokinetics of a Single Dose of Oral Meloxicam in Brown Pelicans (*Pelecanus occidentalis*) *Molly Horgan* | | **2657** Plasma Concentrations of Oxytetracycline After Multiple Subcutaneous Injections in Three Sea Turtle Species *Charles Innis* | **2828** Distal Femoral Fracture Repair in a Domestic Ferret (*Mustela putorius furo*) *Lauren Schmidt* |
| 8:55-9:00 AM | **Break to Move to Next Session** | | | | |
| 9:00-9:15 AM | **2931** Diagnosis of Seminoma with Metastasis to the Brain & Kidneys Using Advanced Contrast Based Imaging *Nick Kirk* | **2932** Pharmacokinetics of the Antimalarial Drug Primaquine in African Penguins *Jessica Carag* | **2791** The Herptile Egg: Staging & Diagnosis for the Reptile & Amphibian Practitioner *J. Jill Heatley* | **2681** Disease Processes Associated with Propagation for Reintroduction of Threatened Sierra Nevada Frogs *Michael Garner* | **2750** The First Report of Successfully Surgical Anastomosis of Cecocolonic Stenosis with End-to-Side Technique in Rat *Chaowaphan Yinharnmingmongkol* |
| 9:15-9:20 AM | **Break to Move to Next Session** | | | **Break to Move to Next Session** | |
| 9:20-9:35 AM | **2941** Concurrent Parosteal Osteosarcoma & Myositis Ossificans in a Fischer's Lovebird (*Agapornis fischeri*) *Athena Gianopoulos* | **2735** Pharmacokinetics & Safety of Amantadine in Orange-winged Amazon Parrots (*Amazona amazonica*) *Kyra Berg* | | **2789** Agreement Between Three Methods to Determine Blood Testosterone Concentrations in Two *Cyclura* Species *Sean Perry* | **2887** Adenomatous Hyperplasia of the Rete Ovarii in a Chinchilla *Federica Fogli* |
| 9:35-9:50 AM | **Break to Move to Next Session** | | | **Break to Move to Next Session** | |
| 9:50-10:05 AM | **2852** Neoplasia in Small Poultry Flocks in Ontario *Elysha Smith* | **2950** Pharmacokinetics of Selamectin After a Single Transcutaneous Dose in Lorikeets (*Trichoglossus spp.*) *Juliette Raulic* | | **2816** A Standardized Approach Developing Cryopreservation Protocols for Reptile Semen, the Red Eared Slider, a Model *Sean Perry* | **2844** Salivary Gland Adenitis with Abscessation due to Pseudomonas Luteola in Domestic Ferrets (*Mustela putorius furo*) *Lauren Schmidt* |
| 10:05-10:45 AM | **Exhibit Hall Break** | | | | |

## Thursday, October 3 (cont.)

| | AAV Track 1 | AAV Track 2 | ARAV Track 1 | ARAV Track 2 | AEMV Track |
|---|---|---|---|---|---|
| LOCATION | PARK VIEW | GRAND B | GRAND C | REGENCY C | REGENCY A & B |
| SPONSOR | | | | | |
| | | | | | **Anesthesia & Analgesia** |
| 10:45-11:00 AM | **2605** The Lumpy Bird: Avian Oncology *Shangzhe Xie* | **2915** 20-Year Trend in References Used for Avian Drug Dosages in an Exotic Animal Formulary *Nicola di Girolamo* | **2845** Pathology Lesions of Reptile, Amphibian, & Fish Eggs & Embryos *Drury Reavill* | **2800** Investigating the Use of Shed Skin for Detection of *Ophidiomyces ophiodiicola* in Snakes *Bryce Dibadj* | **2738** Evaluation of Sedative & Post-Sedative Effects of Alfaxalone-Midazolam & Ketamine-Midazolam in African Pygmy Hedgehogs *Shawna Hawkins* |
| 11:00-11:05 AM | | **Break to Move to Next Session** | | **Break to Move to Next Session** | |
| 11:05-11:20 AM | | **2867** Stomach Morphology of the Budgerigar (*Melopsittacus undulatus*) with Reference to *Macrorhabdus* *Rob Marshall* | | **2819** Hematologic Analysis of Wild Lake Erie Watersnakes & Effect of Sex, Site & Ophidiomycosis Status *Brina Gartlan* | **2743** Cardiovascular Effects of Increasing Doses of Norepinephrine on Isoflurane-Induced Hypotension in Healthy Rabbits *Olivia Uccello* |
| 11:20-11:25 AM | | **Break to Move to Next Session** | | **Break to Move to Next Session** | |
| 11:25-11:40 AM | | **2853** Function of the Intermediate Zone: Unrecognized Third Part of the Psittacine Stomach *Rob Marhsall* | | **2833** Health Status Assessment of Traumatic Injury in Freshwater Turtles *Alison Savo* | **2779** Evaluation of Sedative Effects of Alfaxalone-Hydromorphone versus Ketamine-Midazolam-Hydromorphone for Intravenous Catheterization in the Domestic Rat *Andreanne Rondeau* |
| 11:40 AM-1:00 PM | **Lunch On Your Own** | | | | |
| | **Clinical Cases** | **Diagnostics & Anesthesia** | | **Clincial Research** | **Opthalmology & Dermatology** |
| 1:00-1:15 PM | **2840** Neurofibroma as the Cause of Unexplained Wing Droops in the Red-Tailed Hawk *David Scott* | **2661** Application of Capillary Electrophoresis as a Research & Diagnostic Tool in Avian Medicine *Carolyn Cray* | | **2700** Nitrate Toxicity in Pet Australian Freshwater Turtles *Anne Fowler* | **2669** Management of Glaucoma in Rabbits *Daniela Yuschenkoff* |
| 1:15-1:20 PM | **Break to Move to Next Session** | | | | |
| 1:20-1:35 PM | **2608** Casque Squamous Cell Carcinoma Excision & 3D-Printed Replacement in a Great Pied Hornbill (*Buceros bicornis*) *Shangzhe Xie* | **2786** Validation of Activated Clotting Time Using Diatomaceous Earth in Double-Crested Cormorants *Kyle Abbott* | | **2664** Detection of Nidoviruses, *Mycoplasma*, & *Chlamydia* in Pythons *Rachel Marschang* | **2644** Bilateral Anterior Eye Chamber Abscess Associated with *Encephalitozoon cuniculi* Infection in a Two-Month-Old Guinea Pig *Monica Sidorowicz* |
| 1:35-1:40 PM | **Break to Move to Next Session** | | | | |
| 1:40-1:55 PM | **2647** Mucus Accumulation of the Ventral Air Pouch in a *Leptoptilos crumeniferus* *Ciro Cococcetta* | **2945** Development of Optimal Anticoagulant & Centrifugation Protocol for Preparation of Thrombocyte-Leukocyte Rich Plasma in the Muscovy Duck (*Cairina moschata domestics*): A Pilot Study *Krista Keller* | | **2630** Snakes in the Suite: Husbandry & Veterinary Care of the Ball Python in Biocontainment Laboratories *Lynn Miller* | **2709** Presurgical Stereotactic Radiotherapy to Reduce Tumour Volume in a Suspected Injection-related Sarcoma in a Rabbit *Ellen Risidi* |
| 1:55-2:00 PM | **Break to Move to Next Session** | | | | |

REGISTRATION – REGENCY COAT ROOM – OPENS 7:30 AM

EXHIBIT HALL – REGENCY BALLROOM – 9:00 AM - 5:00 PM

# THURSDAY, OCTOBER 3

*Please note, this grid is subject to change slightly as speaker conflicts arise.*
*A copy of the completed grid will be provided at time of registration on-site.*

REGISTRATION – REGENCY COAT ROOM – CLOSES 3:00 PM

EXHIBIT HALL – REGENCY BALLROOM – 9:00 AM - 5:00 PM

| | AAV Track 1 | AAV Track 2 | ARAV Track 1 | ARAV Track 2 | AEMV Track |
|---|---|---|---|---|---|
| LOCATION | PARK VIEW | GRAND B | GRAND C | REGENCY C | REGENCY A & B |
| SPONSOR | | | | | |
| | | | | | **Dermatology** |
| 2:00-2:15 PM | **2650**<br>Surgical Treatment of Chronic Throat Inflammation in a Marabou (*Leptoptilos crumeniferus*)<br>*Markus Büker-Steinschulte* | **2934**<br>Comparison of Live & Terminal Intravascular Contrast Agents in a Domestic Goose (*Anser anser domesticus*)<br>*Nick Kirk* | | **2809**<br>Snapping Turtle Tips, Tricks & Insights<br>*Renee Schott* | **2817**<br>Treatment of Epitheliotropic T-Cell Lymphoma in a Lop Rabbit<br>*Jenilee Morrison* |
| 2:15-2:20 PM | **Break to Move to Next Session** | | | | |
| 2:20-2:35 PM | **2826**<br>A Novel Treatment of Exuberant Granulation Tissue in an Umbrella Cockatoo (*Cacatua alba*)<br>*Kelly Rockwell* | **2790**<br>Fluoroscopic Angiography in Hispaniolan Amazon parrots (*Amazona ventralis*)<br>*Jamie Douglas* | | **2886**<br>Biotinylation of Reptilian Erythrocytes<br>*Anke Stoehr* | **2783**<br>Dermal Anatrichosomiasis in a Domestic Ferret<br>*Louisa Asseo* |
| 2:35-2:40 PM | **Break to Move to Next Session** | | | | |
| 2:40-2:55 PM | **2878**<br>Multifarious Avian Inuries & Surgical Management<br>*Ushma Patel* | **2850**<br>Evaluation of the Effects of Midazolam on Food Intake in Dudgerigars (*Melopsittacus undulatus*)<br>*Grayson Doss* | | **2787**<br>The Reptile CBC: Investigating Differences Between & Variability of Leukocyte Quantification Methods<br>*John Winter* | **2780**<br>A Case of Cutaneous Botryomycosis in a Domestic Rabbit (*Oryctolagus cuniculus*)<br>*Dominic Byron-Chance* |
| 2:55-3:25 PM | **Exhibit Hall Break** | | | | |
| | | | | | **Cardiovascular & Neurology** |
| 3:25-3:40 PM | **2609**<br>Diagnosis & Treatment of 27 Cases of Proventricular Dilatation Disease (PDD)<br>*Guoliang Tang* | **2674**<br>Sedation in Birds<br>*Grayson Doss* | | **2920**<br>Evaluation of Two Treatment Options for *Cryptosporidium serpentis* Infection in Asymptomatic Eastern Indigo Snakes<br>*James Bogan* | **2688**<br>Pericardial Effusion & Myocarditis in a Guinea Pig (*Cavia porcellus*) with Thromboemboli<br>*Laura Neuhaus* |
| 3:40-3:45 PM | **Break to Move to Next Session** | | **Break to Move to Next Session** | | |
| 3:45-4:00 PM | **2794**<br>Diagnosis & Management of Crop Dilation & Subcutaneous Herniation in Two Avian Species<br>*Rosa Banuelos* | | | **2767**<br>Poison Dart Frogs (*Dendrobates tinctorious*) & Leopard Frogs (*Lithobates pipiens*) Exhibit Stress Response to Restraint<br>*Connor Jacobson* | **2693**<br>Clinical & Pathological Findings in Rabbits with Cardiovascular Disease: 59 Cases (2000-2018)<br>*Sarah Ozawa* |
| 4:00-4:05 PM | **Break to Move to Next Session** | | **Break to Move to Next Session** | | |
| 4:05-4:20 PM | **2615**<br>Intestinal Herniation in a Blue-fronted Amazon (*Amazona aestiva*)<br>*Stephanie Lamb* | | | **2830**<br>Sleeping Slytherin! Evaluating the Efficacy of Alfaxalone in Corn Snakes<br>*Kelly Rockwell* | **2865**<br>Naturally Occurring Tetanus in a Domestic Rabbit (*Oryctolagus cuniculus*)<br>*Joao Brandao* |
| 4:20-4:25 PM | **Break to Move to Next Session** | | **Break to Move to Next Session** | | |

## Thursday, October 3 (cont.)

| | AAV Track 1 | AAV Track 2 | ARAV Track 1 | ARAV Track 2 | AEMV Track |
|---|---|---|---|---|---|
| LOCATION | PARK VIEW | GRAND B | GRAND C | REGENCY C | REGENCY A & B |
| SPONSOR | | | | | |
| | | | | | **Cardiovascular & Neurology** |
| 4:25-4:40 PM | **2729** Management of Chronic Cloacitis in an African Penguin (*Spheniscus demersus*) *Jennilee Morrison* | **2842** Radiographic Heart Size in Wild Raptor Species *David Scott* | | **2724** Specific Anatomy & Contrast Radiography of the Gastrointestinal Tract in Inland Bearded Dragons (*Pogona Vitticeps*) *Karina Mathes* | **2822** Cervical Myelopathy Due to Atlantoaxial Subluxation with Absence of the Dens in a Domestic Rabbit *Kelsey Chapman* |
| 4:40-4:45 PM | **Break to Move to Next Session** | | | | |
| 4:45-5:00 PM | **2607** Successful Diagnosis/ Management of Iron Storage Disease in a Blue & Gold Macaw (*Ara Araruana*) *Ellen Rasidi* | **2933** Grey Parrot Upper Respiratory Tract Imaging & Anatomy *Nick Kirk* | | **2776** Normal Ocular Parameters in Healthy Adult Bearded Dragons (*Pogona vitticeps*) *Krista Keller* | **2771** Handling a Prickly Situation: Characterizing the Neurologic Examination in African Pygmy Hedgehogs *Grayson Doss* |

REGISTRATION – CLOSES 3:00 PM

EXHIBIT HALL – 9:00 AM - 5:00 PM

**Thank you for attending ExoticsCon 2019!**
**We'll see you at ExoticsCon 2020 in**
**Denver, CO, August 29 - September 3!**

Pagination in this book matches the original digital media

# Prevalence of Herpesviruses in Chelonians in Veterinary Practice in Europe

Rachel E. Marschang,
Ekaterina Kolesnik,
Elisabeth Müller

**Session #2839**

*Affiliation:* From Laboklin GmbH & Co. KG, Steubenstr. 4, 97688 Bad Kissingen, Germany. *Key words:* *Testudo*, tortoise, turtle, testudinid herpesvirus, TeHV

*Note:* Abstract format selected by author(s). See Front Matter for description of conference proceedings manuscripts submitted as scientific abstracts.

*Abstract:* Herpesviruses are important pathogens in chelonians. A variety of strains have been detected in various species, including the testudinid herpesviruses (TeHV) 1-4 in tortoises. TeHV 1 and 3 are most commonly found in tortoises in Europe, while TeHV4 has only been described in a very few cases in African tortoise species, and TeHV2 has only been found in *Gopherus* spp. in the United States. A variety of other herpesvirus strains has been found in aquatic turtles. Season is hypothesized to play a role in the epidemiology of herpesviral disease and shedding. This study retrospectively examined herpesvirus detection in samples submitted to a commercial veterinary laboratory for virologic testing between 2014 and 2018. Virus detection was carried out using three different PCRs targeting the herpesviral DNA-dependent-DNA-polymerase gene. Of 3250 chelonians tested during that period, 257 (7.9%) were herpesvirus positive. TeHV1 was detected in 123 (47.5%) cases, TeHV2 in 1 (0.4%) case, TeHV3 in 122 (47.1%) cases, and TeHV4 was detected in 3 (1.2%) cases. Herpesviruses also were detected in 10 emydid turtles. Evaluation of the affected species showed that, among tortoises, Russian tortoises (*Testudo horsfieldii*) were most often affected (20.0% positive), followed by radiated tortoises (*Astrochelys radiata*) (14.1% positive), and spurred tortoises (*Centrochelys sulcata*) (11.7% positive). The highest number of emydid turtle herpesviruses were found in box turtles (7 out of 25 cases, 28% positive). Evaluation of the detection rates according to season showed that TeHV3 was most often found in the spring and summer, while TeHV1 showed no clear seasonality.

# Identification of a Novel *Actinomyces* sp. in Granulomatous Cloacal and Hemipenal Inflammation in Captive Male Ball Pythons (*Python regius*)

Robert J. Ossiboff, DVM, PhD, DACVP,
Steven B. Tillis,
Marley E. Iredale, DVM,
April L. Childress,
James F.X. Wellehan, DVM, PhD, Dipl ACZM, Dipl ACVM (Virology), Dipl ACVM (Bacteriology and Mycology),
Ramiro Isaza, DVM, MS, MPH, Dipl ACZM

**Session #2881**

*Affiliation:* From the Department of Comparative, Diagnostic, and Population Medicine (Ossiboff, Tillis, Iredale, Childress, Wellehan) and the Department of Small Animal Clinical Sciences (Isaza), College of Veterinary Medicine, University of Florida, 2015 SW 16th Ave, Gainesville, FL 32610, USA.

In a large ball python breeding colony, a unique syndrome of granulomatous inflammation of the cloaca and hemipenes was observed in 139 of 481 (29%) breeding males, but absent in virgin males (n = 199). In contrast, only 1 out of 1446 breeding females and no virgin females (n = 187) exhibited similar cloacal lesions. On post-mortem examination (n = 13, 12 males, 1 female), there were numerous pericloacal, hemipenal, and pericoelomic granulomas; fewer animals exhibited intracoelomic and/or intrahepatic granulomas. Histopathologic examination revealed chronic granulomatous inflammation of the affected tissues associated with mixed morphology bacteria. Aerobic culture of a hepatic granuloma of one snake resulted in the isolation of a fastidious, gram-positive bacilli; amplification and sequencing of the 18S gene of the isolate and phylogenetic analysis identified the bacterium as a novel *Actinomyces* sp. A specific, heminested PCR developed to screen for the bacterium within the tissues of affected snakes confirmed the presence of the agent in granulomas of all 13 snakes, as well as cloacal swabs taken at the time of necropsy in 11/13 snakes. To determine if the bacterium was present in unaffected snakes, cloacal swabs (n = 74) of breeding and virgin males and females. The novel *Actinomyces* sp. was identified in swabs from both sexes, with 86% of males (n = 22) and 57% of breeder females (n = 23) testing PCR positive. This study characterizes a novel disease syndrome of breeding male ball pythons associated with a novel bacterium. While the bacterium can be detected in females, it was only very rarely associated with clinical disease.

# Detection of New Nidoviruses and an Orthoreovirus in Association with Respiratory Disease in Veiled Chameleons

Anke C Stöhr, DVM,
Laura Hoon-Hanks, DVM, DACVP,
Amanda Anderson, DVM,
Dawn Evans, DVM, DACVP,
Javier Nevarez, DVM, Dipl ACZM, DipECZM (Herpetology)

**Session #2903**

*Affiliation:* From the Department of Veterinary Clinical Sciences, Louisiana State University, Skip Bertman Drive, Baton Rouge, LA 70803, USA, the Department of Microbiology, Immunology, and Pathology, Colorado State University, 1682 Campus Delivery, Fort Collins, CO 80523, USA, the Louisiana Animal Disease Diagnostic Laboratory, River Road, Baton Rouge, LA 70803, USA.

*Note:* Abstract format selected by author(s). See Front Matter for description of conference proceedings manuscripts submitted as scientific abstracts.

*Abstract:* During the last decade, new viruses have been identified in association with respiratory disease in reptiles, especially snakes, but disease etiology and epidemiology remain poorly understood. Increased mortality associated with respiratory signs occurred in a collection of juvenile veiled chameleons (*Chamaeleo calyptratus*). The animals were part of a research project and were obtained from different breeders in the United States. Two animals were euthanized for diagnostics. On histopathology multifocal catarrhal and lymphocytic rhinitis, and heterophilic and proliferative pneumonia with abundant mucus were found. *Salmonella* spp. were cultured from pooled liver and lung samples. PCR for detection of ophidian paramyxovirus was negative. Sanger sequencing was performed on lung tissues from these individuals and one naturally deceased animal. Full coding genomes of a new nidovirus and an orthoreovirus, as well as partial genomic sequences of a closely related nidovirus were obtained. All viruses were distinct from previously detected reptilian viruses, but showed highest nucleotide identity values to nidoviruses found in snakes and shingleback lizards (69-74% identity) and an orthoreovirus detected in a green bush viper (75% identity) on partial sequences including one of the most conserved genomic regions (RNA-dependent RNA polymerase). Primers for detection of these nidoviruses were designed and clinically healthy subadult and adult veiled chameleons at the research facility tested positive via oral swab by PCR. This is the first description of these viruses in chameleons. Although the primary pathogenicity of each single virus remains unclear, the concurrent infections may have favored clinical respiratory disease in juvenile, likely incomplete immunocompetent animals.

# The Subcarapacial Vessel in Chelonians: Safe for Intravenous Injections?

Kelly Rockwell, DVM,
Nathalie Rademacher, DVM, Dipl ACVR, Dipl ECVDI,
Michelle Osborn, PhD,
Javier G. Nevarez, DVM, PhD, Dipl ACZM, DipECZM (Herpetology)

**Session #2838**

Affiliation: From Louisiana State University School of Veterinary Medicine, Skip Bertman Drive, Baton Rouge, LA 70803, USA.

*Note:* Abstract format selected by author(s). See Front Matter for description of conference proceedings manuscripts submitted as scientific abstracts.

*Abstract:* Chelonian patients are often sampled or administered medications in the subcarapacial vessel. The safety of this site has been questioned due to reports of adverse effects associated with this procedure due to proximity to the cervical vertebrae. The purpose of this study was to characterize distribution of contrast media administered in the subcarapacial sinus and assess for extravascularization. This study protocol was approved by the Louisiana State University IACUC committee. Twenty juvenile red-footed tortoises (*Chelonoidis carbonaria*) were sedated with alfaxalone (15-20 mg $kg^{-1}$ IM) for computed tomography (CT) scans. Preliminary scans were performed on each animal with the head retracted and extended. Then, 1 ml/kg of iodinated non-ionic contrast media was administered in the subcarapacial vessel and scans repeated immediately after administration and at five minutes post contrast. The results of the CT scans revealed that 14/20 (70%) of tortoises had contrast in the trachea, 16/20 (80%) had contrast in the lung, and 14/20 (70%) had contrast in the subarachnoid space. Based on these results, administration of medications in the subcarapacial vessel of red-footed tortoises should be done with caution due to the risk of extravascularization into the respiratory tract and subarachnoid space. This risk may also be found in other chelonians with a more convex carapace.

# Indications, Outcomes and Complications of Prefemoral Reproductive Surgery in Client-Owned Chelonians: 22 Cases, 2017-2018

Nicola Di Girolamo, DMV, MSc(EBHC), PhD, DipECZM(Herp)

**Session #2917**

*Affiliation:* From Avian, Exotic and Zoo Medicine, Center for Veterinary Health Sciences, Oklahoma State University, Stillwater, OK 74078, USA and Tai Wai Small Animal and Exotic Hospital, Hong Kong, Hong Kong.

*Abstract:* This case series includes all client-owned female chelonians that underwent reproductive tract surgery between May 01, 2017 and July 28, 2018 in a mixed exotic and small animal hospital by a board-certified reptile specialist. A total of 22 chelonians underwent surgery. The technique employed in this case series resulted in successful surgeries.

## Introduction

Since the development of prefemoral celiotomy in chelonians, this technique has been widely taught and used in clinical practice and research. Currently, there are no studies assessing the outcomes of surgery in client-owned reptiles, presented for treatment. The purpose of the present case series is to describe outcomes and complications of the use of a modified surgical approach to coeliotomy in chelonians.

## Methods

This case series includes all client-owned female chelonians that underwent reproductive tract surgery between May 01, 2017 and July 28, 2018 in a mixed exotic and small animal hospital by a board-certified reptile specialist. Reproductive tract surgery was defined as surgery involving oviducts, ovaries or both. All female chelonians that underwent any type of reproductive tract surgery in the study period and for which surgery was performed were included.

## Results

A total of 22 chelonians underwent surgery. The following procedures were performed: ovariectomy, ovariectomy and oviductal egg removal, ovariectomy, oviductal egg removal and deep cloacal egg removal, oviduct prolapse removal and ovariectomy, removal of multiple eggs from the urinary bladder and ovariectomy. Of those, 1 animal was euthanized during surgery, 1 animal died during recovery. The remaining 20 animals (90.9%) recovered from the surgery and were discharged.

## Conclusions

There is a wide variety of reproductive surgeries that may need to be performed during clinical activity. The technique employed in this case series resulted in successful surgeries, with an acceptable prognosis.

# Effect of Lithium Heparin Versus Ethylenediaminetetraacetic Acid on the Hematology of Ball Pythons

Lauren Schmidt, DVM,
Melinda Camus, DVM, DACVP,
Laila Proença, MV, DVM, MS, PhD, Dipl ACZM

**Session #2866**

*Affiliation:* From Oklahoma State University, Stillwater, OK 74078, USA (Schmidt), VCA West Los Angeles Animal Hospital, Los Angeles, CA 90025, USA (Schmidt, Proenca), University of Georgia College of Veterinary Medicine, Department of Pathology, Athens, GA 30602, USA (Camus).

*Note:* Abstract format selected by author(s). See Front Matter for description of conference proceedings manuscripts submitted as scientific abstracts.

*Abstract:* Hematology is routinely used in assessing animal health. Previous studies comparing the effect of anticoagulants on hematologic values in non-mammalian species have shown variability in reported hemolysis, packed cell volume (PCV), and thrombocyte aggregation.[1-4] Ball pythons are commonly kept as companion animals and studies evaluating the best anticoagulant for this species are lacking. We compared the effects of lithium heparin (Sarstedt™ Micro tube 1.3 ml lithium heparin) and dipotassium ethylenediaminetetraacetic acid (EDTA; Greiner Bio-One™ MiniCollect® TUBE 0.5 ml K 2E K 2 EDTA) on hematologic values of ball pythons. Blood samples were collected via cardiocentesis from 10 adult snakes and equal aliquots were placed in tubes containing EDTA or heparin. The ETDA- and heparin-preserved samples were then stored in tubes (BD Microtainer™ Capillary Blood Collector and BD Microgard™ Closure) with formalin and submitted for CBC. A blood smear without anticoagulant was submitted for leukocyte differential count. The PCV and total solids concentrations between both anticoagulants were significantly different, with greater values found for EDTA. There was visible hemolysis in all of the heparinized samples, but rarely in EDTA samples. Thrombocytes were more often reported as clumped or decreased in heparinized samples as compared to EDTA-preserved samples, which were all reported as adequate. Based on the results of the present study, samples preserved in EDTA seem to be superior to heparin in determining hematologic values in ball pythons; however, further studies with a larger number of samples are needed before EDTA can be recommended as the first choice anticoagulant in ball pythons.

## References

1. Guzman DS, Mitchell MA, Gaunt SD, et al. Comparison of hematologic values in blood samples with lithium heparin or dipotassium ethylenediaminetetraacetic acid anticoagulants in Hispaniolan Amazon parrots (*Amazona ventralis*). *J Avian Med Surg.* 2008;22:108-113.

2. Hanley CS, Hernandez Divers, S, Bush S, Latimer KS. Comparison of the effect of dipotassium ethylenediaminetetraacetic acid and lithium heparin on hematologic values in the green iguana (*Iguana iguana*). *J Zoo Wildl Med.* 2004;35:328-332.

3. Harr KE, Raskin RE, Heard DJ. Temporal effects of 3 commonly used anticoagulants on hematologic and biochemical variables in blood samples from macaws and Burmese pythons. *Vet Clin Pathol.* 2005;34: 383-388.

4. Muro J, Cuenca R, Pastor J, et al. Effects of lithium heparin and tripotassium EDTA on hematologic values of Hermann's tortoises (*Testudo hermanni*). *J Zoo Wildl Med.* 1998;9:40-44.

# A Twenty-Year Trend in Type of References Used for Reptile Drug Dosages in an Exotic Animal Formulary

Marianne Caron, BS,
Cassie Smead, BS,
João Brandão, LMV, MS, DipECZM (Avian), Nicola Di Girolamo, DMV, MSc, PhD, DipECZM (Herpetology)

**Session #2914**

*Affiliation:* From the Department of Veterinary Clinical Sciences, Center for Veterinary Health Sciences, Oklahoma State University, 2065 W. Farm Rd, Stillwater, OK 74078-2041, USA.

*Note:* Abstract format selected by author(s). See Front Matter for description of conference proceedings manuscripts submitted as scientific abstracts.

*Abstract:* Formularies are likely the most frequently used literature resources in clinical practice. Our aim was to evaluate the type of references that support dosages in a popular exotic animal formulary and its trend over more than 20 years. As a part of a larger project including different animal taxa, data related to the reptile taxa were extracted from 5 editions of an exotic animal formulary. For this subset of the project, only references cited in tables related to "treatments" were included. Each reference was individually categorized. In total, 1,302 citations were evaluated. After removal of references not cited in appropriate sections, 1,089 citations remained. Of those, 580 derived from journal articles (53.2%; 95% CI: 50.3-56.2%), 257 from conference proceedings (23.6%; 21.1-26.2%), 156 from edited books (14.3%; 12.4-16.5%), and 61 (5.6%; 4.4-7.1) from other sources, including other formularies, monographs, magazines, and personal communications. Of the potential primary sources (journal articles and conference proceedings, n = 823), 334 were literature reviews (31.5%; 28.8-34.4%), 191 pharmacokinetic studies (18.0%; 15.8-20.4%), 105 case series (9.9%; 8.2-11.9%), 85 randomized and observational studies (8.0%; 6.5-9.8%) and 57 case reports (5.4%; 4.2-6.9%). There was a significant increasing trend in primary sources from the first (34.1%) to the fifth edition (53.6%) (Odds ratio: 0.45; 95% CI: 0.27-0.73%; $P < 0.001$). Overall, there was an improvement over time in the type of references cited; however, clinicians should still carefully scrutinize the reference section of the formulary whenever prescribing a drug since almost half of the dosages are currently extrapolated from secondary sources.

# A Twenty-Year Trend in Type of References Used for Reptile Reference Intervals in an Exotic Animal Formulary

Cassie Smead, BS,
Marianne Caron, BS,
João Brandão, LMV, MS, DipECZM (Avian), Nicola Di Girolamo, DMV, MSc, PhD, DipECZM (Herpetology)

**Session #2919**

*Affiliation*: From the Center for Veterinary Health Sciences, Oklahoma State University, 001 Boren Veterinary Medical Hospital, Stillwater, OK 74078-2041, USA.

*Note:* Abstract format selected by author(s). See Front Matter for description of conference proceedings manuscripts submitted as scientific abstracts.

*Abstract:* Formularies are among the most frequently used literature resources in clinical practice and are often used as a quick resource to reference intervals. Our aim was to evaluate the type of references that support reference intervals in a popular exotic animal formulary and its trend over more than 20 years. As a part of a larger project including different animal taxa, data related to the reptile taxa were extracted from 5 editions of a drug formulary. For this subset of the project, only references cited in the "reference interval" tables of the formulary were included. Each reference was individually categorized. In total, 1,302 citations were evaluated. After removal of references not cited in appropriate sections, 168 remained. Of those, 107 derived from journal articles (63.7%; 95% CI: 56.2-70.6%), 24 from edited books, 19 from conference proceedings, and 12 from monographs. Of the potential primary sources (journal articles and conference proceedings, n = 126), 98 were reference interval studies (77.8%; 69.8-94.1%), 13 literature reviews, and 15 other primary sources. There was an increase of reference interval studies from the first to the fifth edition from 54.5-63.6%. Overall, there was an improvement over time of types of references cited; however, clinicians should still carefully scrutinize the reference section of the formulary since around one-third of the citations to reference intervals in the current edition of the formulary do not cite primary research.

**Gap in pagination due to formatting issues.**

**Page 66**

# Effectiveness of Common Disinfecting Agents Against Isolates of *Nannizziopsis guarroi*

Caitlyn Hemby,
Krista A. Keller, DVM, Dipl ACZM,
David Sanchez-Migallon Guzman, LV, MS, DipECZM (Avian) DipECZM (Small Mammal), Dipl ACZM,
Joanne Paul-Murphy, DVM, Dipl ACZM, Dipl ACAW,
Barbara A. Byrne, DVM, PhD, Dipl ACVIM, Dipl ACVM,
Daniel Bruce Raudabaugh, MS,
Andrew N. Miller, MS, PhD,
Matthew C. Allender, DVM, MS, PhD, Dipl ACZM

**Session #2893**

*Affiliation:* From University of Illinois College of Veterinary Medicine, 1008 W Hazelwood Dr Urbana IL 61802, USA (Hemby, Keller, Raudabaugh), University of California Davis, 1 Shields Ave, Davis, CA 95616, USA (Guzman, Paul-Murphy, Byrne, Miller, Allender).

*Note:* Abstract format selected by author(s). See Front Matter for description of conference proceedings manuscripts submitted as scientific abstracts.

*Abstract: Nannizziopsis guarroi*, a keratinophilic fungus previously reported as *Chrysosporium guarroi* and/or *Chrysosporium* anamorph of *N. vriesii*, is an emerging infectious disease in companion reptiles. Treatment of affected lizards has often associated with recrudescence of infection, suggesting ineffective treatment or environmental persistence allowing for reinfection. Effective disinfection is needed to prevent contamination of surfaces and equipment used in the care of these animals. To this aim, the qualitative in vitro disinfecting capability of 8 commonly used household and laboratory disinfectants (Novalsan®, 3% bleach, 10% bleach, Virkon®-S, Lysol® household cleaner with hydrogen peroxide, 70% ethanol, 409® and household cleaning ammonia) was tested at 2 different contact times (2 and 10 minutes) and with 3 different aqueous concentrations of 4 molecularly confirmed isolates of *N. guarroi*. A positive control after contact with saline also was grown. After contact with disinfectant or saline, the isolates were plated on potato dextrose agar and incubated for 10 days at 21°C under 12 hours light/dark. The only disinfectant that completely inhibited growth for all 4 isolates regardless of concentration was 10% bleach, at both contact times. Most disinfectants (Virkon®, 3% bleach, Lysol® household cleaner, 70% ethanol, Novalsan® and 409®) exhibited variable inhibition of fungal growth that was isolate dependent. All 4 isolates grew after contact with ammonia, regardless of concentration or contact time. In conclusion, a minimum of 2-minutes exposure to 10% bleach is recommended for disinfection of surfaces and instruments contaminated with *N. guarroi*.

# Treatment of a Retrobulbar Adenocarcinoma in a Diamond Python (*Morelia spilota spilota*) Using Surgery and External Beam Radiation

Joshua Llinas, BVSc (Hons), MVS, BSc

**Session #2621**

*Affiliation:* From Greencross Vets, Brisbane, Queensland, Australia.

Neoplasia is frequently reported in reptiles with the highest incidence reported in snakes.[1] Surgery is the most frequently used mode of treatment in reptiles with reported complications including recurrence of the cancer at the surgery site, metastatic disease, and infection.[1-3] In mammalian veterinary patients, chemotherapy and radiation therapy are increasingly being used when clean margins cannot be achieved, or dissemination of the cancer has already occurred. Radiation therapy has been used in a limited number of reptile cases,[1,4] although there are cases of success being recorded, including a single case of radiation therapy used to successfully treat a periorbital mass in a central bearded dragon *Pogona vitticeps*.[5] A case of retro-bulbar adenocarcinoma in a captive diamond python (*Morelia spilota spilota*) treated with a combination of surgery and radiation therapy is described below.

A spectaculectomy followed by enucleation and resection of grossly abnormal retrobulbar tissue, was performed. Histopathology of the retrobulbar tissue confirmed it to be adenocarcinoma. As clean margins could be not achieved, definitive external beam radiation therapy was offered to the client and the client consented. A successful treatment of a periocular adenocarcinoma in a central bearded dragon (*Pogona vitticeps*) using 10 doses of 4.8 Gy for a total dose of 48.0 Gy was reported by Bishop et al. and was the basis for the prescribed treatment of 10 fractions of 4.8 Gy for a total of 48 Gy delivered on a Monday, Wednesday, Friday schedule to the patient.[5] No direct negative effects were observed and there was no visible evidence of regrowth observed. Two months post treatment, the snake was observed moving abnormally. Full body radiographs showed significant vertebral osteopathy. Inability to control pain resulted in the patient being humanely euthanized. The patient's head was submitted for histopathologic review which confirmed the presence of adenocarcinoma in the previous surgery site. Although treatment was unsuccessful, this case provides information to clinicians that may help develop future protocols using external beam radiation for treatment of neoplasia unable to be removed entirely by surgery alone.

## References


1. Christman J, Devau M, Wilson-Robles H, et al. Oncology of reptiles: diseases, diagnosis, and treatment. *Vet Clin North Am Exot Anim Pract.* 2017;20(1):87-110.

2. Frye FL. Diagnosis and surgical treatment of reptilian neoplasms with a compilation of cases 1966-1993. *In Vivo.* 1994;8(5):885-892.

3. Steffey MA. Principles and applications of surgical oncology in exotic animals. *Vet Clin North Am Exot Anim Pract.* 2017;20(1):235-254.

4. Kent MS. Principles and applications of radiation therapy in exotic animals. *Vet Clin North Am Exot Anim Pract.* 2017;20(1):255-270.

5. Bishop T, Cumming B. Treatment of a squamous cell carcinoma in a bearded dragon (*Pogona vitticeps*) using radiation therapy. *Proc Assoc Avian Vet (AC) Unusual Exot Pet.* 2014.

# Septic Arthritis in a Bearded Dragon (*Pogona vitticeps*) Caused by *Mycobacterium* spp.

Julián Vaillard, MVZ,
Laura A. B. Aguilar, MVZ, EMCV,
Adriana Méndez Bernal MVZ, MSc,
Isaac Martínez Racine, MVZ, MSc,
Ricardo Walter Czaplewski Cícero, MVZ, EMCV,
Aurora Ramos Garduño, MVZ, MSc

**Session #2795**

*Affiliation:* From the Department of Ethology and Wildlife (Vaillard, Aguilar, Czaplewski, Ramos) and the Department of Pathology (Méndez, Martínez), Veterinary College, National Autonomous University of Mexico, Av. Universidad 3000, Coyoacan, Ciudad Universitaria, 04510, Mexico City, Mexico.

*Note:* Abstract format selected by author(s). See Front Matter for description of conference proceedings manuscripts submitted as scientific abstracts.

*Abstract*: A 3-year-old female bearded dragon (*Pogona vitticeps*) was presented to the Veterinary Teaching Hospital at the National Autonomous University of Mexico with a history of right hindlimb swelling. Radiographs revealed lysis of the right stifle and right tarsal bone. A fine-needle joint aspirate from the stifle revealed high amounts of macrophages and giant cells, no bacteria were seen. Ziehl-Neelsen stain revealed acid-fast bacilli-like bacteria. Culture was made on blood agar and MacConkey agar, with negative results. Pain management with meloxicam (0.2 mg/kg PO q24h) and tramadol (11 mg/kg PO q12h) was established; the owner declined antimicrobial therapy. Three weeks later, radiographs revealed the progression of the lytic lesions of the distal femur, proximal tibia, and tarsal bones. One month later, the patient showed left hindlimb swelling. Follow-up radiographs revealed a total loss of right stifle as well as tibiotarsal and tarsometatarsal joints; new lytic lesions appeared at the left stifle and tibiotarsal joints; euthanasia was proposed. At the necropsy, granuloma-like lesions were found at the stifles, coxofemoral, tibiotarsal and tarsometatarsal joints; imprints made from these lesions revealed acid-fast bacilli conglomerates, no other gross lesions were identified. Histopathology results described severe granulomatous osteomyelitis and myositis as well as a severe histiocytic and heterophilic hepatitis with acid-fast intrahistiocytic bacilli. PCR results identified these bacteria as a non-tuberculous mycobacteria. To our knowledge, this is the first post-mortem report of septic arthritis associated with *Mycobacterium* spp. in a bearded dragon, only 2 other cases have been reported, both without post-mortem findings.

# Successful Management of Anemia in a Bearded Dragon using Multiple Whole-Blood Transfusions

Laura A. B. Aguilar, MVZ,
Spencer P. Kehoe, DVM,
Jessica R. Comolli, DVM,
Stephen J. Divers, DVM, DipECZM, Dipl ACZM

**Session #2798**

*Affiliation:* From the Department of Small Animal Medicine and Surgery, University of Georgia Veterinary Medical Center, 2200 College Station Road, Athens, GA 30602, USA.

*Note:* Abstract format selected by author(s). See Front Matter for description of conference proceedings manuscripts submitted as scientific abstracts.

*Abstract:* A 4-year-old, female bearded dragon was presented to Emergency Service after two-day history of lethargy, inappetence, and abdominal distention. On physical examination, the patient was quiet, alert and responsive and her mouth was pale. She was hyponatremic (133.8 mmol/L) and severely anemic (PCV/TS: 3%, 3.0 g/dl). Full body radiographs revealed decreased serosal detail and a fluid-line present within the coelom, suggestive of coelomic effusion. The coelomic ultrasound revealed a large volume of coelomic effusion and numerous variably sized follicles. Fluid was obtained from the coelom, the cytology was consistent with acute hemorrhage and egg-yolk coelomitis. Whole-body CT revealed gravid uterus. On day 2, the patient received two whole blood transfusions administrated over four hours via an intra-osseous catheter. On day 4, the PCV/TS were 7% and 2.8 g/dl. The patient received a third whole blood transfusion intra-operatively administrated over 2 hours. Exploratory coeliotomy and ovariectomy were performed. The source of bleeding was found to be coming from the left ovarian pedicle at its attachment to the renal vessels. Two more whole blood transfusions were administrated after surgery. On day 6, the patient looked brighter and the PCV/TS gone up to 17% and 3.0 g/dl. One month later, the patient was bright, and alert and the PCV/TS had risen to 30% and 5.2 g/dl. No negative health effects were noted in follow up. This is the first report describing the successful administration of five whole blood transfusions from different donors over a short period of time, to a bearded dragon with marked anemia.

# Adenocarcinoma of the Membrana Nictitans in an Inland Bearded Dragon (*Pogona vitticeps*)

Lauren Schmidt, DVM,
Grant Rezabek, MPH, DVM,
Erin Stayton, DVM,
João Brandão, LMV, MS, DipECZM (Avian)

**Session #2829**

*Affiliation:* From the Zoological Medicine Service, Center for Veterinary Health Sciences, Oklahoma State University, Stillwater, OK 74074, USA (Schmidt, Brandão), Oklahoma Animal Disease Diagnostic Laboratory, Oklahoma State University, Stillwater, OK 74074, USA (Rezabek Stayton).

*Note:* Abstract format selected by author(s). See Front Matter for description of conference proceedings manuscripts submitted as scientific abstracts.

*Abstract:* A 9-year-old, intact male, inland bearded dragon (*Pogona vitticeps*) presented for a 3-month history of soft tissue swelling over the left eye. On presentation, a pink multi-lobulated soft tissue mass extended from the medial canthus of the left palpebrae. There was moderate serous discharge from the site with scant hemorrhage from the mass when agitated. Otherwise, the patient had multiple irregularly angled digits and a mildly kyphotic posture but ambulated appropriately. The orthopedic abnormalities were suspected to be related to nutritional or renal secondary hyperparathyroidism. Euthanasia was elected due to the chronicity of the lesion, and a suspected neoplastic process associated with the proliferative periocular tissue, which has been reported previously in this species.[1-4] Necropsy and histopathology were performed. Grossly there was an ulcerated, exophytic, irregular 1 × 1 cm mass on the left nictitans that appeared to infiltrate into the cranioventral periorbital tissue. Histopathology revealed the conjunctival surface was expanded by a papillary, exophytic proliferation of simple columnar to haphazardly arranged and stacking glandular epithelium. Findings were consistent with a well-differentiated adenocarcinoma. Periocular adenocarcinomas have been previously reported in bearded dragons, although not specifically associated with the membrana nictitans.[1] Nictitating membrane adenocarcinoma should be considered in evaluation of periocular lesions in bearded dragons.

## References

1. Darrow B, Johnstone McLean N, Russman S, Schiller C. Periorbital adenocarcinoma in a bearded dragon (*Pogona vitticeps*). *Vet Ophthalmol.* 2013;16:177-182.

2. Gardhouse S, Eshar D, Lee-Chow B, et al. Diagnosis and treatment of a periocular myxosarcoma in a bearded dragon (*Pogona vitticeps*). *Can Vet J.* 2014;55(7):663-666.

3. Hannon DE, Garner MM, Reavill DR. Squamous cell carcinomas in inland bearded dragons (*Pogona vitticeps*). *J Herpetol Med Surg.* 2011;21(4):101-106.

4. Pryor S, Cutler D, Yau W, Diehl K. Adnexal cystadenoma in a bearded dragon (*Pogona vitticeps*). *J Exot Pet Med.* 2018;27(2):85-89.

# Diagnosis and Treatment of Proliferative Fungal (*Fusarium proliferatum*) Stomatitis in a Bearded Dragon (*Pogona vitticeps*)

Rosa M. Bañuelos, DVM,
Shachar Malka, DVM, DABVP (Avian Practice),
Caitlin M. Renna, DVM,
Nicolle T. White, VMD,
Michael M. Garner, DVM, DACVP

**Session #2792**

*Affiliation:* From Long Island Bird and Exotics Veterinary Clinic, 333 Great Neck Rd, Great Neck, NY 11021, USA (Bañuelos, Malka, Renna, White), Northwest ZooPath, 654 W. Main Street, Monroe, WA 98272, USA (Garner).

*Note:* Abstract format selected by author(s). See Front Matter for description of conference proceedings manuscripts submitted as scientific abstracts.

*Abstract:* A 7-year-old male bearded dragon presented to an exotics veterinary clinic with an 18-month history of proliferative stomatitis. Approximately 12 months after initial visualization of lesions, the patient presented to a veterinarian, who initiated oral enrofloxacin therapy, administered at 10 mg/kg q24h for 10 days. This initial therapy decreased swelling and bleeding, but over the next 6 months after this treatment, the lesions worsened, becoming severely erythematous, enlarged, and bleeding. The patient was then presented to an exotics veterinary clinic, where the proliferative lesions were surgically resected, and the tissue was submitted for histopathologic, culture, and sensitivity evaluations. Histopathology showed intralesional bacterial and mycotic components, and fungal culture revealed growth of *Fusarium proliferatum*, a fungal plant pathogen, which previously has been found to cause keratitis and cellulitis in humans. Voriconazole was administered orally (10 mg/kg PO q24h for 45 days; compounded at Wedgewood Pharmacy, Swedesboro, NJ, USA) in conjunction with ceftazidime, selected to treat the cultured intralesional *Pseudomonas aeruginosa*. Ceftazidime (20 mg/kg IM q72h; Tazicef, Ceftazidime for Injection USP; Hospira Inc. Lake Forest IL, USA) was administered for a total of 6 weeks. The patient's oral cavity healed from surgery without incident, and at the time of this writing, has displayed a mild recurrence in clinical signs after discontinuation of voriconazole treatment. To the author's knowledge, this is the first report of *Fusarium proliferatum* in reptilians. Infection with *Fusarium proliferatum* should be considered as a differential diagnosis in cases of reptilian patients with proliferative stomatitis lesions.

# Don't Poke the Dragon: Tail Base Aneurysm in an Inland Bearded Dragon (*Pogona vitticeps*)

Nicole Furst, BA,
Amy Alexander, DVM, Dipl ACZM,
Rachel Turner, DVM,
Robert Ossiboff, DVM, PhD, DACVP,
Simon Swift, Vet MB, Cert SAC RCVS, Dipl ECVIM-CA (Cardiology),
Penny Regier, DVM, MS, Dipl ACVS,
Erin Porter, DVM, Dipl ACVR,
Robson Giglio, MV, PhD, Dipl ACVR,
Leslie Charles, DVM,
James F.X. Wellehan, DVM, PhD, Dipl ACZM, Dipl ACVM (Virology), Dipl ACVM (Bacteriology and Mycology)

**Session #2722**

*Affiliation:* From the Department of Comparative, Diagnostics, and Population Medicine (Alexander, Turner, Ossiboff, Charles, Wellehan) and the Department of Small Animal Clinical Sciences (Swift, Regier, Porter, Giglio), College of Veterinary Medicine, University of Florida, 2015 SW 16th Ave, Gainesville, FL 32610-0126, USA.

*Note:* Abstract format selected by author(s). See Front Matter for description of conference proceedings manuscripts submitted as scientific abstracts.

*Abstract:* A 9-year-old male bearded dragon (*Pogona vitticeps*) was presented for a bulge on the ventrum of the tail base. Ultrasound revealed a hypoechoic structure, which yielded frank blood. Doppler interrogation confirmed it was an aneurysm by demonstrating pulsatile jets into the lumen. Bloodwork was unremarkable, and a microsporidia PCR of a cloacal swab was negative. For three months, the animal's appetite and attitude remained normal with conservative management, but he ceased to defecate. Then, the animal was presented for acute regurgitation and tenesmus. Coelomic ultrasound showed a distended stomach and proximal small intestinal intussusception. During an exploratory coeliotomy, the ventral abdominal vein and mesenteric vasculature were tortuous and swollen. The colon and cecum were severely distended, and blood clots were in the coelom. Humane euthanasia was elected. Necropsy revealed a tail base aneurysm, presumably arterial, associated with vascular thrombosis and fibrinonecrotizing vasculitis of the vent and tail vasculature. The walls of the aneurysm exhibited osseous metaplasia and smooth muscle replacement by fibrous connective tissue. No infectious organisms were seen. Aneurysms have been observed on the neck and in the thoracic cavity of bearded dragons.[1,2] To the authors' knowledge, this is the first report of an aneurysm in the tail of a bearded dragon. The cause of this disease is unknown, but trauma, hypertension, and genetic predisposition have been suggested.[3] The microsporidian *Encephalitozoon pogonae* has been associated with fatal hemopericardium and aneurysms in bearded dragons.[4] This case report provides information on an important differential for swellings on bearded dragons.

*Acknowledgments:* We thank the doctors, technicians, staff, and veterinary students at the University of Florida, College of Veterinary Medicine for outstanding care of this animal.

## References


1. Barten SL, Wyneken J, Mader DR, et al. Aneurysm in the dorsolateral neck of two bearded dragons (*Pogona vitticeps*). *Annu Conf Assoc Reptilian Amphib Vet.* 2006;43-44.

2. Sweet C, Linnetz E, Golden E, Mayer J. 2009. What is your diagnosis? *J Am Vet Med Assoc.* 234(10):1259-1260.

3. Raiti P. Husbandry, diseases and veterinary care of the bearded dragon, *Pogona vitticeps*. *Annu Conf Assoc Reptilian Amphib Vet.* 2011;36-48.

4. Wünschmann A, Armién AG, Childress AL, Wellehan JFX, et al. Intrapericardial *Encephalitozoon pogonae*-associated arteritis with fatal hemopericardium in two juvenile inland bearded dragons. *J Vet Diagn Invest.* 2019. In press. http://doi.org/10.1177/1040638719834330.

# Surgical Correction of Urinary Bladder Prolapse in a Burmese Star Tortoise

Chaowaphan Yinharnmingmongkol, DVM, and
Warinda Somrtith, DVM

**Session #2753**

*Affiliation:* From Animal Space Hospital, 141/1 Puttamonthon Sai 2 Road, Salathummasop, Thaweewuttana, Bangkok, 10170, Thailand.

*Note:* Abstract format selected by author(s). See Front Matter for description of conference proceedings manuscripts submitted as scientific abstracts.

*Abstract:* A 9-year-old, 2.13-kg, female Burmese star tortoise (*Geochelone platynota*) was presented to the hospital due to urinary bladder prolapse. According to its history, the tortoise was observed with signs of constipation, lethargy, depression and anorexia for a week. Full physical examination revealed a fluid-filled urinary bladder, which prolapsed outside the body. Radiographs revealed 3 calcified eggs and intestinal sand impaction. The urinary bladder prolapse was treated immediately to protect the bladder wall. Celiotomy was performed to correct the urinary position and remove the retained eggs by making a window flap with epoxy putty closure technique.[1] A Lego brick was used as a wheel to prevent the plastron from touching with the floor. This technique provides a faster plastron healing process. Esophagostomy tube placement also was performed to maintain post-operation daily nutritional care.[2,3] Post-operative care included daily wound care. Ceftazidime (Cef-4, Siam Bheasach Co, Ltd, Bangkok, Thailand) at 20 mg/kg q72h and meloxicam (Metacam, Labiana Life Sciences S.A., Barcelona, Spain) at 0.5 mg/kg q48h were used for 2 weeks.[4] With no complications, the tortoise was discharged from the hospital 2 weeks after the operation. At followup 8 months later, the plastron wound was perfectly healed. In conclusion, the technique of using a Lego brick to lift up the plastron wound is recommended in chelonian celiotomy.

## References

1. Lock BA. Reproductive surgery in reptiles. *Vet Clin North Am Exot Anim Pract.* 2000;3(3):733-752.

2. Di Girolamo N, Mans C. Reptile soft tissue surgery. *Vet Clin North Am Exot Anim Pract.* 2016;19(1): 97-131.

3. De Voe, Ryan S. Nutritional support of reptile patients. *Vet Clin North Am Exot Anim Pract.* 2014;17(2): 249-261.

4. Mader D. *Reptile Medicine and Surgery.* 2nd ed. St. Louis, MO: Elsevier Saunders; 2006.

# Urinary Retention Demonstrated by Means of Diagnostic Imaging in a Caiman Lizard (*Dracaena guianensis*)

Michaela Gumpenberger, Dr med vet,
Martina Konecny, Mag med vet,
Denise Thaller, Dr med vet

**Session #2752**

*Affiliation:* From Diagnostic Imaging (Gumpenberger), Clinic for Avians and Reptiles (Konecny), Department of Small Animals and Horses (Thaller), Institute of Pathology and Forensic Medicine, vetmeduni vienna, Veterinärplatz 1, 1210 Vienna, Austria.

A Caiman lizard (*Dracaena guianensis*) was presented with severe obstipation at the clinic. In sonography a large, cystic, urinary-bladder-like structure was detected occupying the caudal half of the coelomic cavity and seemingly blocking the pelvis. A fluid sample was collected and identified as urine. Since the occurrence of a urinary bladder in *Teiidae* has been heavily debated in the scientific community, a CT scan with intravenous contrast media application (2 ml/kg BW, Unilux® 370 mg I/ml, Sanochemia, Austria) was performed in addition. Contrast enhancement could then be traced from the kidneys along the ureters into the cloaca and an assumed urinary bladder.

The patient received a paramedian laparotomy and enterotomy where impacted snail shells were removed to release the obstipation. Furthermore, a severe dilatation of the colon was diagnosed. Despite the patient's speedy recovery, it was reintroduced with a similar condition 5 weeks later. Conservative treatment (warm water baths, oral paraffin oil, enema) could not remedy the obstipation acceptably. That is why another laparotomy was performed, however, the Caiman lizard did not survive the second procedure. Unfortunately, the suspected urinary bladder remained undetected in necropsy.

In the meantime, 5 additional Caiman lizards and 3 black and white tegus (*Salvator merianae*) received CT examinations. Anatomy appeared similar to the previously described patient. In 3 animals, intravenous contrast media was injected and finally pooled in the presumably existent urinary bladder. It is the authors' assumption that this formation might resemble a kind of distended urodeum or cloacal diverticle or vestigial urinary bladder.

Diagnostic imaging was indispensable for learning about the origin of the obstipation and for planning appropriate surgery procedures. Unfortunately, final (histologic) proof for the existence of a urinary bladder could not be provided by the pathologist.

*Note:* Initial results of this work were previously presented at the International Conference on Avian Herpetological and Exotic Mammal medicine (ICARE), 2019. Final results are intended to be published as a case series.

# Necrotizing Fasciitis and Osteomyelitis due to *Fusarium solani* in a Russian tortoise (*Agrionemys horsfieldii*)

Jana Mazor-Thomas, DVM, MS

**Session #2769**

*Affiliation:* From VCA Animal Medical Center, Portsmouth, NH 03801, USA.

*Abstract:* A 2-year-old male Russian tortoise was presented for a deep abscess of the left hind limb. Surgery was performed, but the wound recurred, more aggressively, two months after surgery. Histopathology and PCR confirmed a fungal infection (*Fusarium solani*). Multiple therapies including deep intramuscular wound catheter, surgical debridement, intracoelomic and oral antifungals, and antifungal-impregnated beads were used over 10 months of treatment; as of the time of this writing, the infection appears to have been treated successfully.

## Case Overview

A 2-year-old male Russian tortoise was presented in July of 2018, for a deep abscess of the left hind limb, with possible splinter foreign body in the center of profuse pyogranulomatous reaction. Surgery appeared to be successful, but the wound recurred, more aggressively, 2 months after surgery. A second procedure was performed to debride the wound, explore it, and collect samples for aerobic bacterial culture. At this time, the wound extended along fascial planes of the muscles of the distal and proximal limb; it was debrided again, and open wound management begun. There was, again, improvement in the swelling and discharge after debridement, but the swelling gradually worsened. A CT scan was performed to explore for any bony involvement and to investigate for any remaining foreign material that might be in the deeper wound; none was seen, but osteomyelitis of the left xiphiplastron was found. This lesion was aggressive with marked enlargement over the course of one week despite treatment with oral and injectable antibiotics indicated by the aerobic bacterial culture.

In a third surgical debridement of both the limb and plastron lesions, samples were collected for aerobic and anaerobic bacterial culture, as well as fungal culture. When all cultures were negative, a second set of samples was submitted to a separate lab for acid-fast staining for *Mycobacterium*, and then for histopathology, which confirmed a marked, focally extensive, aggressive necrotizing dermatitis, cellulitis and osteomyelitis with fungal hyphae present. Ribosomal sequencing confirmed the species as *Fusarium solani*.

Over the course of the next 5 months, a wide variety of treatments were attempted to control this fungal infection. These included: further open wound management targeted at fungal treatment; intracoelomic injections of amphotericin B; a deep intramuscular wound soaker catheter; oral voriconazole; and antifungal-impregnated calcium carbonate beads implanted into the wound itself. Ketoprofen was also used to decrease inflammation in the limb. The patient removed four esophageal feeding tubes over the course of treatment, complicating our options for management. Because of the length of time and variable results of fungal cultures, our treatment was partially empiric, with several sets of biopsies taken for monitoring. Each treatment option, along with the patient's response and any noted side effects, will be discussed in the following sections.

By March of 2019, multiple biopsies confirmed negative fungal growth. The patient was continued on oral voriconazole for 28 days following the negative biopsies. As of the time of writing, no evidence of recurrent infection has been seen.

## Therapeutic Options

### Open wound management

After the second procedure, and for most of the following five months, a wide variety of bandaging techniques were employed. Initially a wet-to-dry style bandage with saline was attempted, but because of the patient's predilection for defecating on the wound site, it proved impossible to keep the bandage clean. We then used manuka honey under an absorbent dressing, then covered with tegaderm and vetwrap; this protected the wound from fecal contamination; however, honey is an excellent growth medium for fungal infection, so once the primary infection was determined to be fungal in origin, this was discontinued. Silvadene (silver sulfadiazine) is a common wound dressing, with antifungal properties, but the petroleum base was felt to be potentially irritating to the deep intramuscular tissues exposed by the infection. A water-based silver gel (Silversept) was used instead, with no apparent irritation of the deeper tissues.

### Intracoelomic and intralesional amphotericin B injections

*Fusarium solani* is notorious both for locally aggressive behavior (as in this case), for resistance to most antifungals, and for difficulty in correlating in vitro MIC to clinical response.[1] This was complicated by the general paucity of pharmacokinetic data on most antifungals in chelonians. An MIC was run at the Cornell University Animal Health Diagnostic Center, with the caution that results should be considered for research purposes only. The lowest MICs were for Amphotericin B (2.0 µg/ml) and for voriconazole (1.0 µg/ml); all other results would be interpreted as "resistant" in mammalian species. Because the esophageal feeding tubes were repeatedly removed, for several months of treatment we relied primarily on intracoelomic injections, at 0.5 mg/kg once daily for the first 7 days, then reduced to every other day, to reduce the likelihood of renal or hepatic toxicity. Each intracoelomic injection was diluted in 5% dextrose in water, according to manufacturer's instructions, then in 12 ml of reptile Ringer's solution to assist in maintaining hydration. The plastron lesion responded fully and rapidly to this treatment, but the distal limb lesions did not. We suspect that the plastron lesion was exposed to slightly higher and more reliable concentrations, due to its proximity to the coelomic space; while vascular injury to the limb may have prevented sufficient amounts from reaching the most severely affected tissue. It is also possible that long-term administration of q24h amphotericin, rather than decreasing to every other day, would have been more efficacious.

Early on in treatment, wound soaker catheters were also placed in the deepest intramuscular lesions and used to instill 0.25 mg/kg amphotericin B in a more local fashion. This amphotericin was also reconstituted in 5% dextrose in water, according to manufacturer's instructions, but not diluted further. This treatment appeared empirically effective in the local area but caused muscular necrosis and suspected lymphatic obstruction to the rest of the leg. Though amphotericin B has been anecdotally used in SC injections in other species, it is not recommended (especially in the more concentrated form) for deep intramuscular use.

### Oral voriconazole

As mentioned above, MIC data were suggestive that voriconazole might also be an effective treatment. In the initial phase of treatment, a dose of 10 mg/kg PO q24h was used, to reduce the risk of hepatic injury from multiple concurrent medications. The patient's size, and the placement of esophageal feeding tubes, made monitoring of blood chemistry impractical. This dose appeared subjectively to be effective in reducing the amount of discharge from the wound, but biopsy of the site confirmed continued fungal growth. After the last feeding tube was placed, the medication was restarted, but at 10 mg/kg q12h PO, as the risk of continued fungal infection was felt to be higher than the risk of hepatic injury. There was intermittent inappetence at this dose, but no sustained weight loss; the twice daily dosing continued for a total of four months without clinical evidence of significant side effects.[2] One chemistry panel obtained during the course of treatment did show normal hepatic enzymes.

### Amphotericin- and voriconazole-impregnated calcium carbonate beads

Because the patient removed multiple esophageal feeding tubes, and intracoelomic amphotericin B appeared to be ineffective for the distal leg lesions, we also used calcium carbonate beads (Kerrier Co, http://kerrier.com). These beads can be made using a variety of medications, and slowly dissolve at the site of implantation, allowing high concentrations of the medication at the target site with minimal systemic absorption. The beads themselves are radiopaque, but fully dissolve after implantation, so further surgery to remove them is not required. These beads appeared to dissolve at standard environmental ranges for Russian tortoises, and clinically appeared to provide local control, but did not appear sufficient to eradicate the infection fully. No local tissue irritation or other side effects were seen.

### Anti-inflammatory therapy

Meloxicam at 0.3 mg/kg PO q24h was given after the first surgery, but after the follow-up procedures, this was changed to ketoprofen at 2 mg/kg IM q48h. Ketoprofen is a non-selective COX inhibitor that may have better control of upregulated prostaglandin receptors in inflamed tissue, though this has not been verified experimentally (M. Corcoran, personal communication).[3] The q48h dosing schedule was chosen based on the limited pharmacokinetic data available in iguanas.[4]

## References


1. Alastruey-Izquierdo A, Cuenca-Estrella M, Monzon A, et al. Antifungal susceptibility profile of clinical *Fusarium* spp. isolates identified by molecular methods. *J Antimicrob Chemother* 2008;61:805-809.

2. Innis CJ, Young D, Wetzlich S, et al. Plasma concentrations and safety assessment of voriconazole in red-eared slider turtles (*Trachemys scripta elegans*) after single and multiple subcutaneous injections. *J Herpetol Med Surg*. 2018;24:28-35.

3. Hannon D. Non-steroidal anti-inflammatory drugs (NSAIDs) in reptiles and amphibians: a review. *Proc ExoticsCon*; 2015;589-603.

4. Tuttle AD, Papich M, Lewbart GA, et al. Pharmacokinetics of ketoprofen in the green iguana (*Iguana iguana*) following single intravenous and intramuscular injections. *J Zoo Wildl Med.* 2006;37(4):567-570.

# Successful Diagnosis and Management of a Plastron Eosinophilic Pseudocyst with a Partial Plastronectomy

Sean M. Perry, DVM,
R. Avery Bennett, DVM, MS, Dipl ACVS,
Mark A. Mitchell, DVM, MS, PhD, DipECZM (Herpetology)

**Session #2785**

*Affiliation:* From the Department of Veterinary Clinical Sciences, Louisiana State University, School of Veterinary Medicine, Skip Bertman Drive, Baton Rouge, LA 70803, USA.

A 1.5-year-old African spurred tortoise (*Centrochelys sulcata*) was presented for a focal plastron swelling located at the junction of the abdominal and femoral scutes that had developed over the past 2 months. No historical, husbandry, or nutritional changes were thought to be attributed to the lesion's development. Examination revealed very mild pyramiding of the shell with a focal ovoid extrusion or swelling of the plastron at the junction of the abdominal and femoral scutes measuring 4 cm × 4 cm. Radiographs were performed showing an osseous cyst like structure. Computed tomography was performed with intralesional Omnipaque™ (iohexol) to determine the cysts integrity and for surgical planning. Fine needle aspirate of the lesion demonstrated cystic fluid with some spindle cells and reactive fibroblasts. A biopsy was performed prior to surgery to confirm an intraosseous fibrosing pseudocyst with intralesional hematoma. Surgical debridement and removal of the intraosseous pseudocyst was performed with alfaxalone 20 mg/kg IM and hydromorphone 0.4 mg/kg SC. A partial plastronectomy was performed in the area of the abdominal and femoral scutes and a cyst like structure was removed from the bone. The surgical site was left open to heal by second intention. Postoperative management of the surgical site included wet-to-dry bandages every 2-3 days for 2 weeks. Additionally, topical silversulfadiazine was initiated 2 weeks post operatively. One month post operatively, bandages were changed to duoderm. Healing progressed and the surgically created plastron defect bone filled in over two months followed by continued keratin healing over 6 months. This is the first documented case of a pseudocyst in a chelonian.

# Management of Respiratory Infection in a Reticulated Python (*Python reticulatus*) using an Osmotic Antibiotic Pump

Meera Kumar Music, BVM&S, MRCVS

**Session #2855**

*Affiliation:* From Center for Bird and Exotic Animal Medicine, 11401 NE 195th St NE, Bothell, WA 98011, USA.

*Abstract:* A 19-year-old albino reticulated python (*Python reticulatus*) presented for open mouth gaping and production of white, mucoid material from the oral cavity. Upon exam, marked increase in respiratory rate and effort was observed. Copious discharge from the trachea was identified, sampled, and submitted for aerobic culture. A complete blood count (CBC) was drawn from the ventral tail vein. No other diagnostics were approved by the owner. The patient was otherwise clinically unremarkable. Aerobic culture grew a heavy growth of both *Klebsiella pneumonia* and *Acinetobacter baumanni*. The CBC supported the presence of a bacterial infection (heterophilia with toxic change). The decision was made to move forward with antibiotic treatment, based on culture results. Due to the animal's size and aggression, various treatment options were considered with staff safety and minimizing stress to the patient in mind. An antibiotic osmotic pump was chosen as the treatment option that allowed regular elution of medication without need for repeated handling/restraint of this patient. The python was anesthetized and an Alzet osmotic pump was placed subcutaneously. After 1 month, the patient was re-anestheized and the implant was removed. All clinical symptoms have resolved, and the recheck CBC supported a resolving infection. A chemistry sample obtained post implantation showed elevated uric acid, which returned to normal 1 month later.

## Introduction

A 19-year-old albino reticulated python (*Python reticulatus*) presented for open mouth gaping and production of white, mucoid material from the oral cavity. This animal belongs to a small collection in a local reptile zoo, with minimal staffing. In order to assure staff safety, aggressive, large reptiles are handled only when the zoo owner is on site. This albino, reticulated python is over 14 feet long, and is highly reactive when handled. The snake was symptomatic for a respiratory tract infection with increased respiratory rate and effort for approximately 1 month prior to presentation. In that time, clinical symptoms had progressed to include white, thick mucoid discharge production from the trachea. Prior to antibiotic choices made, an aerobic culture of tracheal debris was submitted, and a CBC also was performed to try to confirm presence of bacterial infection. The culture grew heavy growth of both *Klebsiella pneumonia* and *Acinetobacter baumanni,* and the CBC confirmed the presence of a heterophilia and toxic change, supporting the diagnosis of a bacterial infection. Blood culture, diagnostic imaging (including, but not limited to radiographs, and/or CT), and serum biochemistry also were recommended to gather a more complete understanding of this patient's health status, but no further diagnostics were approved by the owner due to financial concerns. An extensive discussion was then had regarding methods of treatment, including antibiotic injections, nebulization, and oral dosing. Given the infrequency of feeding, oral dosing was ruled out as a treatment option. Nebulizations were not ruled out, but given the setup at the zoo, there was a concern over how to effectively treat the patient with minimal handling. By the same token, antibiotic injections were discussed for empirical treatment, including ceftazidime (WG Critical Care, LLC, Paramus, NJ) and amikacin (Zoetis, Kalamazoo, MI) based on culture results; however, these still required once every three days injections, which would necessitate active restraint and stress to the animal, and danger to the staff.

Use of osmotic antibiotic pumps have increased in the last several years, particularly with snakes, for long term treatment of ongoing infections, and this was chosen as the treatment modality as it allowed for medication administration with minimal stress and handling of the patient.[1-4]

## Materials and Methods

Alzet osmotic pumps (Durect Corporation, Cupertino, CA) were ordered, and amikacin was chosen as the antibiotic to be instilled based on culture results. The 2ML2 implant size was chosen after calculating the amount of amikacin needed to create appropriate plasma level concentrations. The pumps were loaded using aseptic technique per the usage instructions. Once the pumps were confirmed to be appropriately loaded, they were transported in a sterile container to be implanted at the zoo site.

The patient was manually restrained and anesthetized with midazolam (0.5 mg/kg IM; Akorn, Inc, Lake Forest, IL) and ketamine (25 mg/kg IM; Zoetis, Kalamazoo, MI) administered intramuscularly.[5] After approximately 45 minutes, the patient was deeply sedated/minimally responsive, and was loaded from its display enclosure to an off-display room for its procedure. The animal was surgically prepped using chlorhexidine (Vetoquinol USA, Inc, Ft. Worth, TX) and alcohol. Lidocaine (2 mg/kg SQ; VetOne, Boise, ID) was administered to provide local analgesia at the implantation site. A skin incision was made, and blunt subcutaneous (SQ) dissection was performed to create space for the implant. Once the implant was in place, the surgical site was closed with 2-0 Weblon non-absorbable suture (Patterson Veterinary, Devens, MA) using a horizontal mattress pattern. Zookepers noted that with the anesthetic protocol used, the patient took over 48 hours to fully recover, though it was breathing normally on its own. During the 48-hour period of recovery, the animal was minimally responsive.

Twenty-six days later, the patient was re-anesthetized with ketamine (15 mg/kg IM; Zoetis) and midazolam (0.4 mg/kg IM; Akorn Inc). The reduced doses were used to try and mitigate some of the recovery concerns noted after the patient's first procedure. Once the patient was sedated, lidocaine was once again instilled to provide local analgesia in the implant site. The surgical site was re-opened by removing the sutures and blunt dissection. The implant was removed, and a small amount of yellow-brown, serous discharge was seen. The site was flushed thoroughly with NaCl (0.9%), and 2-0 Webmax absorbable suture (Patterson Veterinary, Devens, MA) was used to close the skin. The anesthetic protocol for this procedure still provided appropriate depth of anesthesia for the surgical procedure, but the recovery period was greatly reduced, at about 10 hours post administration of the anesthetics.

A serum chemistry sample was drawn from the ventral tail vein at the same time as removal of the implant, to assess uric acid (UA) levels. These came back as elevated at 22 mg/dl (reference range 3.5-17.4 mg/dl).

One month later, the patient was fully asymptomatic and the surgical site was healing well. The snake was re-anesthetized with the second protocol a final time to assess the surgical site in detail and redraw for CBC and chemistry. The site showed no swelling or discharge, the CBC showed resolution of the toxic change, and significant improvement of the previously noted heterophilia, and the UA had returned to within normal limits, 4.5 mg/dl.

## Results and Discussion

Significant clinical improvement was noted after use of the implants. Moreover, the patient was able to receive one-off, minimal-stress interactions due to ability to implant antibiotics as a depot, and only interact with the animal once monthly. The staff also was kept safe by minimizing need to interact with the snake. This treatment technique has since been used in other animals at this facility, with similar success.

Culture and sensitivity guidance for antibiotic use was crucial in making treatment decisions for this animal. Areas where improvement to the process could occur include where the procedure was performed, although the site offers an abbreviated surgery room, it is possible that some minor contamination of the implant site caused the inflammation seen upon implant removal, though it did seemingly resolve. The UA levels did show an initial increase, but returned to normal within the month, which may indicate a need to monitor this animal's kidney values in the future months and years. Finally, the doses of ketamine/midazolam changed greatly over this animal's treatment time period, showing that even much lower doses created an effective level of predictable sedation, while allowing for faster recovery. This approach to sedation will certainly come in to play as we continue to use these implants as a treatment modality for our larger, more aggressive species.

*Acknowledgments:* I thank John Sykes for all of his help and guidance in treating this animal, and for giving the initial presentation that inspired me to move forward. Without his help, treatment could not have been so successful, and I am deeply appreciative of the time he took to assist me. I am also grateful for the Reptile Zoo of Monroe, WA for allowing me to try a new treatment modality.

## References

1. Clancy M, Newton A, Sykes J. Management of osteomyelitis caused by *Salmonella enterica* subsp. *houtenae* in a Taylor's cantil (*Agkistrodon bilineatus taylori*) using amikacin delivered via osmotic pump. *J Zoo Wildl Med.* 2016;47(2):691-694.

2. Sykes J, Greenacre C. Techniques for drug delivery in reptiles and amphibians. *J Exot Pet Med.* 2006;15(3):210-217.

3. Sykes J, Ramsay E, Schumacher J, et al. Evaluation of an implanted osmotic pump for delivery of amikacin to corn snakes (*Elaphe guttata guttata*). *J Zoo Wildl Med.* 2006;37(3):373-380.

4. Sykes J. Reptile Long-duration Antibiotics: Are There Options? *Proc North Am Vet Comm*. 2017.

5. Carpenter JW. *Exotic Animal Formulary.* 5th ed. St. Louis, MO: Elsevier Saunders; 2018.

**Gap in pagination due to formatting issues.**

**Page 585**

# Techniques for Minimally Invasive Removal of Urinary Stones in Chelonians

Nicola Di Girolamo, DMV, MSc(EBHC), GPCert(ExAP), PhD, DipECZM (Herpetology)

**Session #2875**

*Affiliation*: From the Zoological Medicine Service, Center for Veterinary Health Sciences, Oklahoma State University, Stillwater, OK 74078, USA.

*Abstract:* Urinary stones are a common presentation in chelonians, likely caused by chronic dehydration or improper diet. In the present master class we will discuss techniques that we have developed in the last years for removing urinary stones (either in the urinary bladder or in the cloaca) without performing a plastronotomy.

## Introduction

Urinary stones are a common presentation in terrestrial chelonians, and are believed to be associated to chronic dehydration, improper diet and husbandry, and insufficient physical activity. In a recent study, 5.1% of client-owned chelonians examined at a veterinary teaching hospital were confirmed to have urolithiasis.[1] There are two main locations where urinary stones are found in chelonians, the urinary bladder (left or right lobe) and the cloaca. It is believed that renal and ureteral urolithiasis does not occur since urinary stone formation happens in the urinary bladder, mainly by congregation of uric acid. Stones located in the urinary bladder can be either an incidental finding on survey radiographs or can result in aspecific clinical signs, (decreased appetite, intermittent constipation, swelling of posterior limbs, loss of small amount of urine). Stones located in the cloaca typically result in more serious clinical signs, including complete lack of defecation, straining, swelling of the cloacal area, and anuria.

While urinary bladder stones are rarely a surgical emergency, since they likely have been present for a long time before the diagnosis, cloacal stones need to be treated promptly to restore urinary and fecal output. Depending on the size and the location of the stone, different techniques can be used to remove urinary stones.

## Urinary Bladder Stones

### Endoscopic-assisted techniques

Endoscopic removal of urinary bladder stones has the advantage that it avoids surgery. It may be extremely valuable in the case of uroliths of moderate to small size in species where access to the urinary bladder via cystoscopy is feasible. In the case of larger stones, endoscopic techniques may not be feasible or may not be ideal.

In order to perform cystoscopy, the endoscope is gently inserted in the vent and directed cranially in order to reach the cloaca. The cloaca is a saclike cavity into which the ureters, gonadal ducts, colon, and bladders empty. In some aquatic turtles, paired saclike structures (ie, the accessory bladders) open into the cloaca dorsal to the colon. The endoscope, once in the cloaca, is directed toward the urethral opening, which is located ventral to the rectum. Access to the urethral sphincter is gained by gentle ventral pressure. Warm fluids are infused (1 drop every 3-4 seconds) during the procedure to allow distension of the urethral opening. Careful attention should be paid to avoid overdistension of the bladder in small chelonians. Gentle pressure and fluid infusion are usually sufficient to gain access to the urinary bladder.[2] Depending on the species, the use of rigid versus flexible endoscopes may be beneficial. Chelonians can be placed in dorsal or ventral recumbence.

Two main techniques for endoscopic removal of urinary stones have been performed by the author.

*Endoscopic grasping:* Removal of the entire stone, in those cases where the stone is smaller than the pelvic canal, has been achieved using either a polypectomy loop snare[3] or a 4-wire basket. Using 2.7-mm rigid endoscopes, commercially available polypectomy loop snares permit removal of larger stones than commercially available 4-wire baskets, with the drawback of providing an inferior grip. After the stone is grasped with the instrument, a gentle traction, associated with continuous fluid instillation, has permitted removal of the stones. Care needs to be taken to prevent damage to the urethra.

*Endoscopic lithotripsy:* Uroliths may be fragmented by use of holmium:yttrium-aluminum-garnet (Ho:YAG) lasers. The laser probe can be inserted in the operative channel of a rigid endoscope allowing proper direction of the beam. Laser settings of 12 Watt/1.2 Joule/10 Hertz to 8 Watt/0.7 Joule/12 Hertz have been successfully used to fragment small stones. Large fragments can be removed using the basket; whereas, small fragments can be flushed out with fluid irrigation. Endoscopic lithotripsy may not be feasible or ideal in the case of large stones in large species due to the time needed to fragment them. In a case seen by the author with a urinary stone of approximately 10 cm, 4 hours of lithotripsy had a minimal impact on the size of the stone. Further studies with different lasers or strategies to fragment large stones are needed to determine feasibility in those cases.

## Prefemoral cystotomy

The use of prefemoral cystotomy and destruction of the stone with a mini-driver surgical drill has been successful in removing large and medium-sized stones. Prefemoral fossa celiotomy (syn. prefemoral coeliotomy) is currently considered the surgical access of choice for most urinary and reproductive surgeries. This approach avoids prolonged post-surgical dry-docking, which is necessary following plastron osteotomy and is particularly indicated in species with a relatively small plastron. In the author's experience prefemoral cystotomy is more easily achieved with the chelonian positioned in oblique positioning with the lower side being the one in which the urinary bladder stone is more easily reached. This recumbency allows the stone to gravitate towards the prefemoral incision, making its approach sensibly easier. The skin, subcutaneous tissue, and the transverse and oblique abdominal muscles, and coelomic membrane are transected and the coelomic cavity accessed. The use of ring retractors is extremely useful to enhance access and visibility. Care should be taken to avoid trauma to the urinary bladder upon entering the coelom. Cystocentesis should be performed if the bladder is distended and impairs surgical access to the coelom. The urinary bladder can be marsupialized to the skin. The stone is destroyed with a mini-driver or a rotary drilling tool and Babcock forceps. Depending on the size of the chelonian, urinary bladder thickness is significantly different. Thin urinary bladders are prone to rupture during suture and require extreme care during this process. Amputation of part of the urinary bladder may be the only solution if major damage has occurred during the procedure or as a consequence of the chronic presence of the stone. Closure of the body wall should be performed in 2-3 layers. It may not always be possible to close the coelomic membrane. Closure of the muscles and skin are routine. If a substantial amount of subcutaneous tissue is present, then additional subcutaneous sutures should be considered. The healing times after prefemoral coeliotomy (approximately 4 weeks) are substantially shorter than after a plastron osteotomy (approximately 1 to 2 years).

## Cloacal Stones

Cloacal stones are urinary stones that have been eliminated from the urinary bladder and get trapped in the caudal pelvis. Rarely, they reach the vent and are visible externally. The following procedures should be performed under general and/or intrathecal anesthesia.

Depending on the species and size of the chelonian affected and on the size and shape of the stone, different techniques can be employed for their removal. Currently, the easiest and most effective way to remove these stones consists of a cloacal approach. The shape of caudal scutes of juvenile or adult tortoises of certain species (eg, African-spurred tortoises, *Centrochelys sulcata*) can limit the access to the tail and the cloaca. In those cases, removal of the caudal carapacial scutes increases visibility and accessibility to the stone. The scutes can be removed by performing a straight-line incision with a rotary tool or an orthopedic driver. Care needs to be taken to prevent injury of the skin at the level of its attachment to the carapace. In species that have minimal development of caudal scutes or in young individuals where the caudal scutes are not yet fully developed to cover the tail and the cloaca, this procedure is not needed.

Once appropriate visualization of the tail and the cloaca is achieved, the stone can be visualized placing a retractor (eg, Lone Star retractor) on the rims of the vent. Alternatively, an endoscope can be used to assist the procedure.[5] After the stone is properly visualized, it can be destroyed by means of a rotary drilling tool and Babcock forceps. Care needs to be taken to minimize damage to the cloacal mucosa.

## References

1. Keller KA, Hawkins MG, Weber EP III, et al. Diagnosis and treatment of urolithiasis in client-owned chelonians: 40 cases (1987-2012). *J Am Vet Med Assoc.* 2015;247(6):650-658.

2. Di Girolamo N, Selleri P. Clinical applications of cystoscopy in chelonians. *Vet Clin North Am Exot Anim Pract.* 2015;18(3):507-526.

3. Mans C, Foster JD. Endoscopy-guided ectopic egg removal from the urinary bladder in a leopard tortoise (*Stigmochelys pardalis*). *Can Vet J.* 2014;55(6):569-572.

4. Nardini G, Bielli M, Nicoli S, et al. Endoscopic laser lithotripsy in chelonians: two cases. *Veterinaria* 2014;28:33-37.

5. Mans C, Sladky KK. Endoscopically guided removal of cloacal calculi in three African spurred tortoises (*Geochelone sulcata*). *J Am Vet Med Assoc.* 2012;240(7):869-875.

# Urolithiasis and Surgery of the Lower Urinary Tract in Reptiles

Heather A. Bjornebo, DVM, DABVP (Reptile and Amphibian Practice), CertAqV

**Session #2641**

*Affiliation:* From Arizona Exotic Animal Hospital, Mesa, AZ 85201, USA.

*Abstract:* Urolithiasis is a common presentation in reptile patients, in particular chelonians and iguanas. Utilizing both a thorough review of the available literature and firsthand experience of the author, this master class lecture will discuss proposed pathophysiologies of urolithiasis in reptiles along with discussion of clinical approaches to diagnosis, treatment, and management of the condition including surgical removal. Surgical approaches will include standard coeliotomy and colostomy in saurians and plastronotomy and pre-femoral approaches in chelonians.

## Introduction

Urinary tract disease is well documented in reptiles with the most common presenting complaints being changes in activity, anorexia or decreased appetite, weakness, muscle tremors, constipation, bloat, dyschezia, cloacal tissue prolapse, and weight loss.[1] Pathology of the bladder is also well documented with urolithiasis one of the most commonly seen in reptile patients, particularly in chelonians and green iguanas (*Iguana iguana*).[2-4] Other conditions involving the bladder include cloacal prolapse of the bladder, infectious cystitis, ectopic ova, and neoplasia.[3,5,6] With exception of infectious cystitis, surgery is often indicated for treatment of bladder pathology in reptiles.[7]

### Normal gross anatomy and function

The reptilian lower urinary system consists of the urinary papillae, cloaca, urethra, and bladder or cloacocolon.[7] The renal filtrate originating from the kidneys flows through the ureters as they extend through the peritoneal membrane ending in the urinary papillae. From there, emptying either into the neck of the bladder, in the case of some chelonians, or dorsal aspect of the cloaca, the common excretory chamber of the urinary, reproductive, and gastrointestinal tract.[3,8,9]

The cloaca is comprised of three regions: the urodeum, coprodeum, and proctodeum. The distinctness and relative size of these regions varies considerably by species.[10] With exception of chelonians, the ureters enter the cloaca at the urodeum from the paired urinary papillae.[3,10] Feces from the gastrointestinal tract is excreted into the coprodeum, located cranially. The largest and most caudal region is the proctodeum, which receives products from both urodeum and coprodeum.[10]

The reptilian bladder is connected by a short urethra to the ventral floor of the urodeum near the urinary papillae.[8,9,11,12] All urinary products must enter and exit through this single communication and it is common to find cloacal contaminants within the bladder. The reptilian bladder is not considered sterile.[8,12] The bladder is highly elastic and the walls are comprised of several layers of transitional epithelial cells (urethelium) and a thin layer of smooth muscle. There is no submucosa.[9,13] Not all reptiles possess urinary bladders. The bladder functions as waste storage and in tortoises is a major site of both water and ion storage in periods of drought.[12] The bladder is absent in snakes, crocodilians, and some lizards.[8,9,12] In *Agama agama* and many members of the family

Teiidae the bladder is rudimentary. The bladder is merely a stalk in *Sceloporus grammicus disparulis* and *S. poinsettia*.[9] In those species lacking a bladder, nitrogenous waste is stored in the distal rectum. When present the bladder is located in the caudoventral coelom. Chelonian bladders may have multiple lobes or two or more accessory urinary bladders attached at the neck of the primary bladder dorsal to the pubis. Their bladders can compromise a significant amount of the body cavity.[5,8]

## Urolithiasis

Clinical signs more specific to urolithiasis include hematuria, egg binding, dysuria, hyorexia, reduced fecal output, stranguria or tenesmus, reduced mentation, and hind limb swelling;[3,14] however, most uroliths are occult and detected by palpation on routine examination or detected incidentally when patients present for unrelated complaints such as respiratory tract infections or dog bites.[5,14]

Cystic calculi in reptiles are most commonly comprised of uric acid dihydrate and/or urate salts such as ammonium urate, sodium urate, calcium urate, or potassium urate. Uroliths comprised of other substances, such as struvite, carbonate apatite, and calcium phosphate, are far less common.[4] In a retrospective study of 100 desert tortoises (*Gopherus agassizii*) in 1999, it was shown that most (73%) of uroliths present as solitary stones. This same study also noted a higher incidence of urolithiasis in males (60%) and stones were located 47% of the time in the left bladder lobe, 30% in the right, and 23% on midline.[15,16]

In general, the cause of cystic calculi formation in reptiles is unknown; however, proposed etiologies include excessive dietary protein or oxalates, vitamin A or D deficiency, bacterial cystitis, presence of foreign material within the bladder such as an ectopic egg or suture remnant, chronic hypothermia, and dehydration.[3,17] As discussed previously, urates and urate salts are more likely to form in alkaline urine, which may explain why herbivorous reptiles appear over represented when it comes to urolith formation. Urolith formation in free-ranging chelonians is thought to be in part caused by chronic dehydration leading to increased and prolonged water resorption from the bladder resulting in concentration of urate salts; however, it is unknown how water access plays a role in captive populations as most captive tortoises have more frequent access if not total free access to water sources. Lack of a burrow or other shelter to escape extreme temperatures has also been suggested, but again its role is dubious.[14]

### Diagnosis

The diagnostic approach to any reptile case should always begin with a good history and physical examination. Husbandry should be thoroughly evaluated and compared to that species' natural environment, diet, and behaviors found in the wild as often deficits in meeting the patient's husbandry needs contribute to the onset of disease and a patient's capacity for recovery.[18,19] Diagnosis of urolithiasis is made by palpation and radiography.[5] Additional diagnostic modalities may include urinalysis, computed tomography, and coelomic ultrasound.[14]

Urinary calculi can be located within the urinary bladder or lodged within the cloaca.[14,20] While cloacally located uroliths may be palpated via cloacal examination or visualized via cloacoscopy, the presence of a shell provides some clinical challenges with physical examination of the coelom in chelonians.[20,21] Palpation via the prefemoral fossae can be performed while the patient is held head-up in vertical position with the carapace facing the practitioner. Hands are then placed within both prefemoral fossae for palpation which can be aided by gently rocking the patient to the left and right. It is possible to palpate eggs, cystic calculi, and other coelomic masses using this procedure; however, access is still very limited.[21]

Radiographically, urate stones in other species are radiolucent; however, the presence of complexed potassium and calcium within reptile urate stones make them more opaque and thus usually visible on radiographs.[5] Uroliths are best viewed on dorsoventral views and are generally located within the mid-caudal coelom. They tend to have a thick, laminar appearance caused by repeated deposition of crystalline material on the external surface of the urolith. Computed tomography results typically agree with those of standard radiography, but may be useful in identifying concurrent pathology elsewhere in the patient such as hepatopathies, pulmonary disease, neoplasia, non-calcified eggs, great vessel mineralization, and ascites.[14]

In rare cases where uroliths cannot be detected on radiographs, ultrasound can be utilized; however, ultrasonography is more commonly used in chelonians to aid in urine collection via cystocentesis than in urolith detection.[14]

Urinalysis performed on chelonians with cystic calculi appears to be unrewarding based on the limited data available for both normal and clinically abnormal patients.[16,17] In non-chelonians with cystic calculi urinalysis often reveals hematuria, pyuria, hypocitraturia, bacteriuria, crystalluria, and pH abnormalities.[14]

In a retrospective study of captive tortoises with urolithiasis, hematologic and biochemical abnormalities were rare. In contrast, wild tortoises studied exhibited elevated blood urea nitrogen (BUN) and uric acid concentrations. This discrepancy could in part be due to the low numbers of tortoises tested in the study and that normal values can vary greatly throughout the season or based on an individual's sex, maturity level, nutrition, environment, and husbandry.[14]

**Treatment**

Traditionally, treatment of urolithiasis in reptiles is based on physical removal of the calculi as dissolution of the stones to date has proven unsuccessful.[14,16] Prevention of stone enlargement may be possible based on one small study that involved five African spurred tortoises (*Geochelone sulcata*). This study utilized Er-jin decalculous herb extract on patients with small sized uroliths. Urolith size was monitored every 2 months via radiographs and no size increase was noted. These tortoises all voluntarily excreted their uroliths within 3 weeks to 20 months following treatment; however, given the very small sample size of the study and lack of a non-treatment control group, it is hard to draw conclusions as to the true efficacy of this treatment.

The method chosen for removal of cystic calculi is highly dependent on their location and size. Also important factors that need consideration are the size and physical conformation of the patient. Cloacoliths may be removed manually, with the aid of long forceps, or with the assistance of cloacoscopy when located within the cloaca. Surgical removal of uroliths via cystotomy is usually required for those located within the confines of the bladder. Chelonian patients will require cystotomy via the prefemoral fossa or plastronotomy;[14] however, it may be possible to remove small uroliths by endoscopy or long forceps via the cloaca.[16]

*Cloacoliths:* As stated, cystic calculi within the cloaca, or cloacoliths, may be removed from the anesthetized patient either manually or with forceps with or with the aid of an endoscope. Larger cloacoliths may require fragmentation. A burr or drill bit can be used to assist in the fragmentation. The practitioner should take care to ensure good visualization of the instrument during this procedure to prevent iatrogenic trauma to the cloacal soft tissue. The use of an endoscope improves visualization, decreasing complication risks. Irrigate the cloaca with tap water to further improve visualization and flush away debris while fragmenting the stone. Mild swelling of the cloacal mucosa is expected following this procedure and post-procedural recovery time is typically rapid. If the stone is not accessible or refluxes back into the bladder, surgery is necessary.[20]

*Cystotomy via standard coeliotomy:* In order to gain surgical access to the bladder lumen, coeliotomy is necessary. The over-all approach to coeliotomy in reptile patients varies depending on the type of patient involved. In lizards and crocodilians, coeliotomy to access the bladder may be performed either at ventral midline or perimedianally. A paramedian approach is often used to avoid the large ventral midline vessel. If a ventral midline approach is utilized this vessel can be identified then moved off to the side or ligated with little clinical complications. A midline approach minimizes the need to transect muscle bellies which decreases postoperative pain which is of benefit to the patient during recovery.[22,23] Care should be taken when entering the coelom to avoid puncturing the large colon. Handle the tissues within the coelom carefully as they tend to be more fragile than that of mammals.[19,23,24] After gaining access to the coelom, one should proceed with cystotomy as discussed later in this report.

After access to the coelom is accomplished, cystotomy is performed. For cystotomy in lizards and chelonians underdoing plastronotomy, the thin walled reptilian bladder is easily located. Gently elevate the bladder out of the coelom then isolate it using stay sutures and sterile saline soaked laparotomy sponges or gauze to minimize contamination of urine into the coelom. Cystotomy then proceeds as it would for mammalian patients. Excess urine can be removed via cystocentesis or suction following incision through the bladder wall.[22,23]

Following removal of the urolith, the bladder lumen should be flushed with copious amounts of sterile saline and suctioned if necessary. The bladder wall is closed either in 1-2 layers using fine absorbable monofilament suture on an atraumatic needle in an inverting pattern taking care to apply only minimal traction to prevent tearing of the fragile tissue and then any stay or marsupialization sutures should be removed.[22,23] Recent studies in both ball pythons (*Python regius*) and loggerhead sea turtles (*Caretta caretta*) concluded that a synthetic absorbable monofilament suture material, such as Monocryl, is preferred.[23,25]

Closure of a standard coeliotomy incision can be accomplished in 2 to 3 layers depending on patient size. The coelomic membrane should be closed with fine synthetic absorbable suture on a taper needle in a simple continuous pattern. The skin will be the holding layer and should be closed in an everting pattern with non-absorbable sutures due to the tendency for reptile skin to invert along the incision line. Surgical glue is often utilized to provide additional strength and as a hydrophobic barrier to the incision.[19,23,24] Closure of prefemoral coeliotomy is accomplished utilizing the same techniques and materials as coeliotomy in lizards.[22,23,26,27]

*Cytotomy via plastronotomy:* Chelonian coeliotomy poses more of a challenge due to the presence of a shell. Options include plastronotomy or a prefemoral approach.[23,28] Cystotomy via plastronotomy is considered the treatment of choice for cystic calculi in chelonians, especially for those stones fairly large in diameter.[14,16] In patients with mineralized plastrons, an osteotomy is performed to gain access to the coelom. A trapezoidal incision is made into the abdominal and femoral plates, preventing damage to the pelvic bones, using either an oscillating sagittal saw or rotary cutting tool with circular cutting blade.[22,23] The use of a sagittal saw allows for cutting the corners of the trapezoid without having to over cut.[28] At least three sides of the incision should be beveled at 45 degree angles in order to prevent the bone flap from collapsing into the coelom when it is later replaced.[23] Lateral cuts should be made first, followed by the cranial incision along the abdominal scutes and last the caudal along the femoral scutes. It is also important while performing the osteotomy to irrigate the incision with sterile fluids to prevent overheating of the tissue and to flush away bone dust. Care should be taken not to damage the underlying musculature and to leave the periosteum intact along the caudal incision.[22,28] Once the osteotomy is complete, the flap is elevated away from the underlying abdominal musculature with periosteal elevators, scalpel handle, or scalpel blade and reflected caudally taking care not to dissect away the periosteum from the bone. The flap is then covered with sterile gauze moistened with sterile saline.[22,23,28] Flushing the osteotomy site and coelomic membrane with copious amounts of sterile saline should be completed in order to remove debris prior to continuing into the coelom.[28]

Entrance into the coelom can be gained by a ventral midline incision.[22,23] Care should be taken to avoid damaging two large venous sinuses located lateral of midline; however, these may be ligated if the need arises.[22] Care should be taken as a midline incision may inadvertently progress into the pericardial sac, which should be avoided. An alternative approach is to ligate one of the paired abdominal veins and proceed with a transverse or crescent shaped incision.[28] From here, one should proceed with cystotomy as discussed previously in this manuscript.

For plastronotomy, closure of the coelomic membrane and abdominal musculature is important to avoid formation of adhesions and is achieved utilizing a simple continuous pattern with a fine absorbable monofilament suture material on a taper needle.[22,23] The bone flap should then be gently cleansed of debris with sterile saline and then repositioned applying gentle pressure to allow apposition of the edges.[22] The flap should then be secured either by use of metal plates and screws or layers of epoxy resins.[22,23,28] The latter is more commonly used and preferred. A 5-minute curing epoxy is recommended.[28] Epoxy is layered with either fiberglass mesh sheets or sterile polypropylene gauze.[22,23,28] Polypropylene gauze is preferred to standard gauze or fiberglass by some practitioners as it does not fray.[28] Bone wax, a layer of silver sulfadiazine cream, or Tegaderm Transparent Film Dressing is placed over the osteotomy incisions to prevent epoxy from entering within. It is important to immerse the gauze or fiberglass in the mixed epoxy resin until saturated then apply in 3-4 thin layers over the incisions, allowing it to dry and harden slightly between each layer.[22,23] Applying a small amount of triple antibiotic ointment over the epoxy will take away any remaining stickiness from the patch's surface.[28]

While it remains the gold standard, there are many drawbacks to cystotomy via plastronotomy that must be considered. The procedure itself is long, requiring extensive anesthesia time. Osteotomy is considered fairly painful with recovery and healing tending to be prolonged. Placement of an esophagostomy tube is often warranted for nutritional support immediately following the procedure as many patients become inappetent. Aquatic patients must be dry docked for several weeks following plastronotomy which can interfere with their normal feeding behavior. Postsurgical complications may be serious such as loss of vascularity of the plastronal flap leading to necrosis and bone sequestrum formation, infection, coelomitis, and dehiscence of the bone flap margins.[23,29] Recently, one example of such a complication involved development of a urinary bladder fistula exiting the plastronotomy site in a radiated tortoise. Luckily this complication was able to be surgically corrected with a positive outcome for the patient.[30]

*Cystotomy via prefemoral fossa coeliotomy:* An alternative that can be considered for chelonian patients with relatively small uroliths is prefemoral fossa coeliotomy as an approach to gain access to the bladder.[15,22,23,29,31] This approach is appropriate for patients with a carapacial length over 15 cm and with a urolith no more than twice the length of their prefemoral fossa.[29] This approach is particularly useful in species with relatively small plastrons as seen in many aquatic and semi-aquatic turtles.[23] The most notable potential drawback of the prefemoral approach is decreased visualization of the bladder. Small rents in the bladder wall or other pathology may go undetected at the time of surgery leading to internal pathology and adhesion formation.[15]

For prefemoral fossa coeliotomy, the patient should be oriented in such a way that it allows the best access and visualization of the coelom via the prefemoral fossa. The exact orientation will depend on the species involved.[23] For tortoises, extend the hind limbs caudally and tape them together to allow exposure of the fossa. Patient is then placed in right lateral recumbency with the hind end slightly elevated to allow for visceral to fall away from the incision. The surgeon should then orient the patient tilting them ventrally just enough to allow for best visualization of the left prefemoral fossa. The surgical site is then sterilely prepared and draped.[22,26,27] A craniocaudal incision is made midway between the carapace and plastron starting at the bridge and extending to the thigh.[22] The subcutaneous and muscle layers are each then transected or bluntly dissected.[22,23,26] The coelomic membrane is then incised along the same line, taking care not to inadvertently enter either the bladder or lung. From here the bladder can be located with atraumatic forceps and the cystotomy performed.[22,23,26,27]

It is not possible to fully exteriorize the bladder when performing cystotomy via prefemoral fossa coeliotomy. A portion of the bladder should be elevated to the incision and either marsupialized to the incision or stay sutures placed to prevent contamination of the coelom. The stone then can be retrieved using a sterile table spoon bent at a right angle. Stones that are too large to be removed can be carefully broken apart or crushed into pieces using forceps or drills. Care should be taken to avoid tearing the fragile bladder wall.[22,23]

### Post-surgical care

As discussed previously, the reptile bladder is not sterile and antibiotic therapy is warranted for prevention of infection post cystotomy.[16] In order to guide antibiotic selection, culture should be collected from within the bladder lumen during surgery. In the interim, initial antibiotic therapy should be selected based on the fact that most isolates in reptile patients are gram negative. A practitioner should always take into account the patient's health status and avoid medications that may be contraindicated based on any concurrent pathology. In addition, the route of administration should also be considered when selecting antimicrobials taking into account both ease of administration and skill level of the administrator. Good hydration should be maintained throughout antibiotic therapy. Common first choice antibiotics include amikacin, ceftazidime, enrofloxacin, and trimethoprim-sulfas.[32]

Postoperative surgical wound healing time in standard coeliotomy and prefemoral coeliotomy averages around approximately 3-4 weeks.[22,29] During this time, care should be taken to keep the incision clean and monitor closely for complications as you would in mammals.[22,23] Healing time for plastronotomy, as discussed previously, is prolonged. Healing of the osteotomy site can take as long as 1 to 2 years. The patch itself can be permanently removed after 1 year; however, close monitoring of the patch is recommended. Temporary removal of the patch every 1-3 months during this time will allow for close inspection of the osteotomy site. After examination a fresh patch should be placed until the site has fully healed.[22]

### Prognosis

Prognosis following urolith removal in reptiles is dependent on the procedure performed and presence of concurrent pathology within the patient. Postoperative complications are rare in all procedures discussed, but the practitioner should be aware of their potential. With proper medical and surgical treatment overall prognosis for treatment of urolithiasis is good; however, little data exists on long term outcomes.[23] Urolith recurrence has been documented, yet little data exists when it comes to recurrence frequency. Prevention is difficult in most cases due to lack of clarity in the exact mechanisms involved in their formation, more study in this area is required. Until then, the best recommendation is routine annual examination to monitor patients for recurrence.[16]

## References

1. Gibbons PM. Reptile and amphibian urinary tract medicine: diagnosis and therapy. *Proc Annu Conf Assoc Reptilian Amphib Vet.* 2007;69-86.

2. Zwart P. Renal pathology in reptiles. *Vet Clin North Am Exot Anim Pract.* 2006;9:129-159.

3. Reavill DR. Urinary tract diseases of reptiles. *J Exot Pet Med.* 2010;19(4):280-289.

4. Kolle P, Hoffman R, Wolters M, et al. Cystic calculi in reptiles. *Proc Annu Conf Assoc Reptilian Amphib Vet.* 2001;191-192.

5. Ferrell ST. Reptile urology. *Proc Am Board Vet Pract.* 2012. Available at: http:/www.vin.com/doc/?id=5329666. Accessed December 30, 2016.

6. Miller HA. Urinary diseases of reptiles: pathophysiology and diagnosis. *Semin Avian Exot Pet Med.* 1998;7(2):93-103.

7. Divers SJ, Innis CJ. Urology. In: Divers, SJ, Stahl SJ, eds. *Mader's Reptile and Amphibian Medicine and Surgery.* 3rd ed. St. Louis, MO: Elsevier Saunders;2019:626-648.

8. Wyneken J. Reptilian renal structure and function. *Proc Annu Conf Assoc Reptilian Amphib Vet.* 2013;72-78.

9. Fox H. The urinogenital system of reptiles. The urogenital system of reptiles. In: Gans C, Parsons TS, eds. *Biology of the Reptilia.* New York, NY: Academic Press;1977:1-157.

10. Wyneken J. Master class: anatomy and physiology of the reptilian gastrointestinal tract. *Proc Annu Conf Assoc Reptilian Amphib Vet.* 2011;49-57.

11. Wake MH. The comparative anatomy of the urogenital system – urogenital system of the turtle. In: Marvalee HW. *Hyman's Comparative Vertebrate Anatomy.* 3rd ed. Chicago, IL: University of Chicago Press;1979:598-601.

12. Jacobson ER. Overview of reptile biology, anatomy, and histology. In: Jacobson ER, ed. *Infectious Diseases and Pathology of Reptiles.* Boca Raton, FL: CRC Press;2007:13-14.

13. Augley E, Frye FL. Reptilian renal system. In: Augley E, Frye FL, eds. *Comparative Veterinary Histology with Clinical Correlates.* Boca Raton, LA: CRC Press;2001:144.

14. Keller KA, Hawkins MG, Webber III EPS, et al. Diagnosis and treatment of urolithiasis in client-owned chelonians: 40 cases (1987-2012). *J Am Vet Med Assoc.* 2015;247(6):650-658.

15. Mader DR, Ling GV. Cystic calculi in the California desert tortoise, *Gopherus agassaizii*: evaluation of 100 cases. *Proc Annu Conf Assoc Reptilian Amphib Vet.* 1999;81-82.

16. Mader DR. Calculi: urinary. In: Mader DR, ed. *Reptile Medicine and Surgery.* 2nd ed. St. Louis, MO: Elsevier Saunders;2006:763-771.

17. Wilson H. Reptile emergencies and critical care. *Proc Atlantic Coast Vet Conf.* 2002. Available at: http://www.vin.com/doc/?id=3846115. Accessed on December 30, 2016.

18. Kirchgessner M, Mitchel MA. Chelonians. In: Mitchell MA, Tully TN, eds. *Manual of Exotic Pet Practice.* St. Louis, MO: Elsevier Saunders;2009:207-250.

19. Nevarez J. Lizards. In: Mitchell MA, Tully TN, eds. *Manual of Exotic Pet Practice.* St. Louis, MO: Elsevier Saunders;2009:164-206.

20. Mans C, Sladky KK. Endoscopically guided removal of cloacal calculi in three African spurred tortoises (*Geochelone sulcata*). *J Am Vet Med Assoc.* 2012;240(7):869-875.

21. Divers S. Clinical approach to tortoises and turtles. *Proc ExoticsCon.* 2015;541-551.

22. Mader DR, et al. Surgery: soft tissue, orthopedics, and fracture repair. In: Mader DR. *Reptile Medicine and Surgery.* 2nd ed. St. Louis, MO: Elsevier Saunders;2006:581-597.

23. Girolamo ND, Mans C. Reptile soft tissue surgery. *Vet Clin North Am Exot Anim Pract.* 2016;19(1):97-131.

24. Hernandez-Divers SJ. Reptile coeliotomy: a vital technique to master. *Proc North Am Vet Conf.* 2006; 1614-1618.

25. McFadden MS. Suture materials and suture selection for use in exotic pet surgical procedures. *J Exot Pet Med.* 2011;20(3):173-181.

26. Proenca LM, Divers SJ. Coelioscopic and endoscope-assisted sterilization of chelonians. *Vet Clin North Am Exot Anim Pract.* 2015;18:555-570.

27. Johnson JD, Bjornebo H. Prefemoral ovariosalpingectomy of desert tortoises (*Gopherus agassazii* and *Gopherus morafkai*). *Proc Annu Conf Assoc Reptilian Amphib Vet.* 2013;25.

28. Boyer TH. Chelonian celiotomy. *Proc Western Vet Conf.* 2011. Available at: http://www.vin.com/doc/?id=5183499. Accessed Dec. 30, 2016.

29. Mangone B, Johnson JD. Surgical removal of cystic calculi via the inguinal fossa and other techniques applicable to the approach in the desert tortoise (*Gopherus agassizii*). *Proc Annu Conf Assoc Reptilian Amphib Vet.* 1998;87-88.

30. Cutler D. Urinary bladder fistula via plastronotomy site in a radiated tortoise (*Astrochelys radiata*). *Proc ExoticsCon.* 2016;833.

31. Sladky KK, Mans C. Preface: soft tissue surgery. *Vet Clin North Am Exot Anim Pract.* 2015;19:xiii-xiv.

32. Mader DR. Antibiotic therapy in reptiles. *Proc* Central Vet Conf. 2008. Available at:
http://veterinarycalendar.dvm360.com/antibiotic-therapy-reptiles-proceedings-0.
Accessed December 30, 2016.

# Gross Lesion Recognition for the Reptile and Amphibian Practitioner

Michael M. Garner, DVM, DACVP

**Session #2687**

*Affiliation:* From Northwest ZooPath, 654 West Main Street, Monroe, WA 98272, USA.

*Note:* Abstract format selected by author(s). See Front Matter for description of conference proceedings manuscripts submitted as scientific abstracts.

*Abstract*: Gross lesion recognition, like other imaging modalities, is a bit of an art form. Vast zones of grey may confront the investigator when considering what the lesion is, what it could be, and what to do with it. And nobody is good at all of it. A seasoned livestock pathologist could possibly clear an entire day's necropsies without the help of a single histology slide. A bee keeper would recognize immediately the maggot stages that parasitize his colony. A dog-cat pathologist in a busy private laboratory or a companion animal clinician may not have a clue about gross lesions in livestock or bees. Historically, our attempts at accurately diagnosing disease by gross lesion recognition have been a humbling experience, and that has been the impetus for all further diagnostic specialties. So now that we have put gross pathology in its proper place, why even bother with such an imperfect science? Gross lesions, or lack thereof, are the first visual indication of what may be wrong with the patient. These are the lumps, the effusions, the asymmetrical oddities, the discolorations, the odiferous clues to disease that (hopefully) stimulate a scientific thought process culminating in a list of differential diagnoses. The purpose of this master class is to present images of common and not so common reptile gross lesions in live and necropsy specimens, and in a participatory manner, establish a differential diagnosis and means for establishing a definitive diagnosis.

## Recommended Reading

1. Campbell TW, Ellis C. *Avian and Exotic Animal Hematology and Cytology.* Ames, IA: Iowa State University Press;2007.

2. Latimer KS, Rakich PM. Necropsy examination. In: Ritchie BW, Harrison GJ, Harrison LR, eds. *Avian Medicine: Principles and Applications.* Lake Worth, FL: Wingers Publishing Inc;1994:356-376.

3. Garner MM. An overview of biopsy, cytology and necropsy techniques for clinical veterinarians. In: Aguilar RF, Hernandez SM, Divers SJ, Perpinan D, eds. *Atlas de Medicina de Animales Exoticos,* 2da edición. Buenos Aires: Inter-Medica;2010:437-443.

# Chelonian Total Intravenous Anesthesia

Stefania Scarabelli, DVM, MRCVS, Cert AVP (VA), DECVAA

**Session #2705**

*Affiliation:* From Clinica Veterinaria Modena Sud, Piazza Dei Tintori 1, 41057 Spilamberto, Italy.

*Abstract:* Chelonian anesthesia represents a challenge for the veterinary anesthetist. Cardiovascular and respiratory anatomy and physiology play important roles in anesthetic management; blood shunting, long periods of apnoea, different breathing patterns among species and temperature-dependent metabolism can all influence the anesthetic event, particularly when inhalant agents are used. Total intravenous anesthesia (TIVA) can be an attractive alternative. Advantages of this technique in mammals include quick induction and recovery and no need for anesthetic vaporizer. In reptiles, TIVA could avoid the use of an anesthetic machine thus allowing safe anesthesia outside the hospital environment. Alfaxalone is commonly used in dogs and cats for TIVA and has been already studied as induction agent in chelonian, used both intravenously and intramuscularly. In the author's experience, its use in TIVA for prolonged procedures is advantageous compared with inhalant anesthesia, even if differences among chelonian species must be considered.

## Introduction

Tortoises and turtles are commonly anesthetized for diagnostic and surgical procedures. Anesthesia in these species has been challenging for several reasons: anatomic and physiologic features, diversity among families and species, paucity of pharmacokinetic and pharmacodynamics data about anesthetic and analgesic drugs.

Historically, intramuscular or intravenous sedation and induction followed by maintenance of general anesthesia with inhalants has been the most commonly used protocol in chelonian; however, maintaining a stable anesthetic plan with inhalant agents can be difficult, due to the peculiar cardiovascular and respiratory physiology of these animals. Total intravenous anesthesia (TIVA) can be an attractive alternative.

## Cardiovascular Anatomy and Physiology Relevant for Anesthesia

Noncrocodilian reptiles have a 3-chambered heart with 2 anatomically separated atria and a single anatomically continuous ventricle, which is functionally divided by a muscular ridge into 2 main chambers: the cavum dorsale and the cavum pulmonale, which are comparable in function to the left and right ventricle, respectively. The cavum dorsale, just ventral to the atrioventricular orifices, accepts blood from both the atria and acts as the outflow tract for the aorta. It can be divided into the cavum arteriosum, which receives oxygenated blood from the left atrium, and the cavum venosum, which receives blood from the right atrium and from which both the aortic arches originate. Below the cavum venosum lies the cavum ventrale or cavum pulmonale.[1] The cavum venosum and cavum arteriosum are connected by an interventricular canal that acts as a large ventricular septal defect. In fact, reptiles are able to shunt blood from left to right and vice versa. Shunting has important functions in life but the degree of shunt will affect inhalant anesthesia, particularly uptake and elimination of drugs. Moreover, shunting will affect reliability of pulse-oximetry, inhalants concentration, and capnography readings.[2]

### Renal portal system

Another unique feature of the reptile cardiovascular system is the renal portal system, a system of vessels that collect blood from the caudal part of the body and carry it directly to the kidney and liver. Several authors suggest that drugs should not be injected in the caudal part of the body because they will be readily metabolized and excreted before reaching the site of action. This concept is very controversial and there are works supporting[3,4] and disputing[5] this hypothesis.

## Respiratory Anatomy and Physiology Relevant for Anesthesia

Chelonian are obligate nasal breathers. The glottis is at the base of fleshy tongue, the trachea has complete tracheal rings and bifurcate quite proximally, with differences among species. All chelonians have paired, multichambered lungs dorsally attached to the carapace. They do not have a diaphragm; they have a septum horizontale that separates the lungs from the rest of the coelomic cavity. Chelonian use abdominal, pectoralis, and pelvic muscles for ventilation, with differences reported within species.[6]

Lung volumes, breathing patterns, and control of ventilation have been studied in different species of tortoises and turtles. Glass and colleagues,[7] for example, compared ventilation in *Pelomedusa* and *Testudo*. The turtle showed periods of breath-holding interrupted by episodes of continuous breathing; whereas, the tortoise exhibited single breaths alternated with short periods of breath-holding (Fig 1). This is important to consider when positive pressure ventilation is applied during general anesthesia and if capnography is conducted and end tidal inhalants and blood gas values are measured.

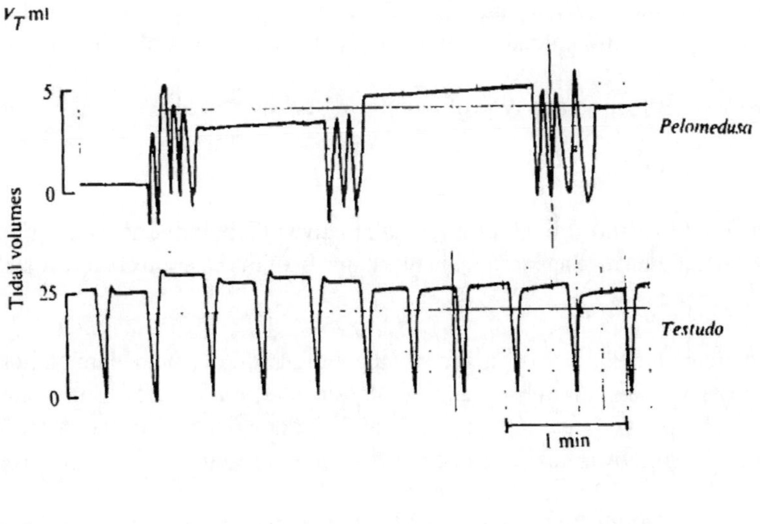


**Figure 1.** Typical patterns of breathing in *Testudo* and *Pelomedusa*.

*Pelomedusa* had a higher mean weight-specific tidal volume compared with *Testudo*; this allows diving species to faster restore gas composition in the lungs after a diving period. In both species, ventilation increased whit hypercapnia; whereas, hyperoxia caused depression of ventilation.

Finally it is worth noting that expiratory time in chelonians is as long as 10 times the inspiratory time[8] and that reptile lungs are considered more fragile compared to mammalian lungs. Thus, peak pressure $< 10\ cmH_2O$ is recommended.[2]

## Temperature and Metabolism

Temperature and metabolism can have an indirect effect on anesthesia. During periods of breath-holding, many reptiles are capable of converting to anaerobic metabolism, which will lead to acidemia. Turtles of the genus *Pseudemys* can survive up to 100% nitrogen for up to 27 hours.[9]

Chelonian are poikilothermic thus depending on environmental temperature. Temperature will influence control of ventilation, cardiac shunting, and metabolism of administered drugs, thus playing a pivotal role on peri-anesthetic management.

## Anesthetic Plan

An anesthetic plan should consider two broad categories: patient related factors and factors related to the procedure that need to be undertaken.

Patient factors should include genus and species, age, weight, husbandry, and physical status. If possible, general anesthesia should be delayed until the patient has received appropriate supportive care. When dealing with unusual species, a literature review should be undertaken.

Procedure-related factors include expected pain (surgical/diagnostic procedure), length of the procedure, location of the procedure (eg, CT cold room), animal positioning, and expected/possible bleeding.

Availability of mechanical ventilation and type of monitoring equipment also should be considered.

### Premedication and induction

The aim of premedication is to avoid stress for the animal, to give preoperative analgesia if required, to decrease the dose of induction and maintenance agents and, in the majority of the cases in chelonians, to allow intravenous catheterization.

Most commonly used premedication protocols include a combination of midazolam and or alpha-2-agonist as sedative, an opioid (most commonly morphine) as an analgesic, and ketamine for further analgesia and immobilization.[10] If high doses of sedatives and ketamine are used, induction of anesthesia is possible too. Higher doses will led to more pronounced cardiovascular and respiratory depression and a prolonged recovery can be expected.

An attractive alternative for short procedures with mild to moderate degrees of pain, is the use of intramuscular (IM) alfaxalone, with or without opioids, non-steroidal anti-inflammatory drugs and local anesthesia. If intravenous access is already available, propofol or alfaxalone can be used as induction agents.

## Maintenance of General Anesthesia

General anesthesia can be maintained with top-ups of the drugs used for premedication, but for prolonged procedures, inhalant anesthesia has been the most commonly used method. Due to the cardiovascular and respiratory physiology of chelonian, inhalant anesthesia can be challenging and unpredictable, with sudden awakening and prolonged recoveries happening relatively commonly compared with other species.[2] For these reasons, total intravenous anesthesia may be a good alternative.

## Total Intravenous Anesthesia

Total intravenous anesthesia (TIVA) refers to induction and maintenance of general anesthesia by intravenous drugs only. In humans, as in domestic species, a hypnotic drug (propofol or alfaxalone) is used in combination with an analgesic, most commonly an opioid. In some cases, the opioid is administered intramuscularly in dogs and cats before induction of general anesthesia and added intraoperatively to the hypnotic agent only if further analgesia is required.[11]

Intermittent boluses could be used for TIVA but it would lead to large variations in drug plasma concentration with higher risk of side effects or inefficacy of the drug itself. For this reason, continuous rate infusions (CRIs) are used that facilitate maintenance of a therapeutic dose (Fig 2).

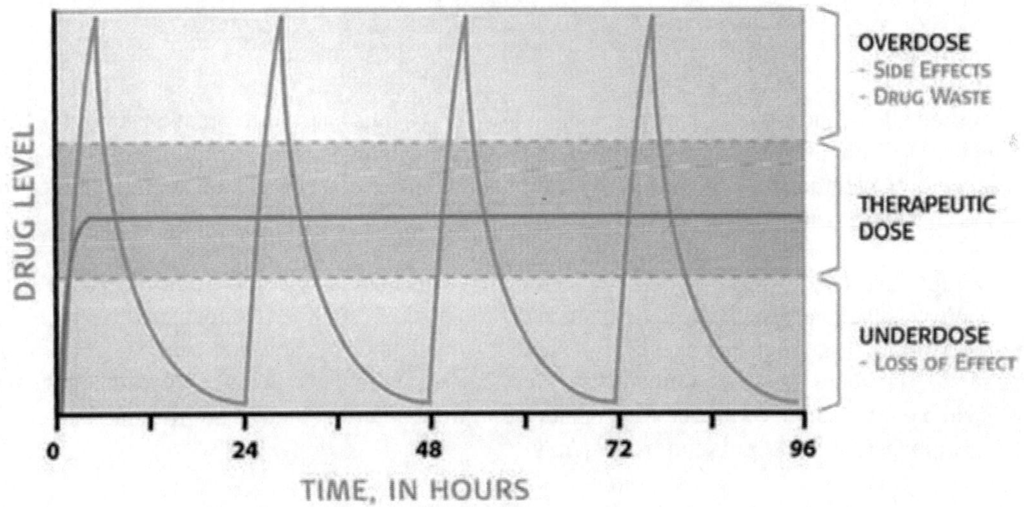


**Figure 2.** Continuous rate infusion drug level over time.

Drugs used for TIVA should have specific pharmacokinetic properties, including rapid onset of action, rapid metabolism, high clearance, and no cumulative effect. Advantages of TIVA reported in human and veterinary medicine include: reduced pollution in the operating theatre, quick induction and recovery, reduced post-operative nausea and vomiting, and no need for anesthetic vaporizer.[12] In mammals, oxygen supplementation and endotracheal intubation (ETI) should be performed; however, in reptiles this may be unnecessary, thus rendering ETI and the availability of a breathing system optional and allowing safer procedures outside the hospital environment.

### Alfaxalone

Alfaxalone is a synthetic neuroactive steroid that works on gamma aminobutyric acid (GABA) receptors, producing hypnosis and muscle relaxation; it does not have analgesic properties. In dogs and cats it is mainly metabolized in the liver via glucoronidation and it is not associated with accumulation.[2,12] It is licensed in dogs, cats, and rabbits for induction of general anesthesia and can be used intravenously (IV) or intramuscularly (IM). Alfaxalone has been successfully used for TIVA in dogs[13] and cats[14] as well as for prolonged periods.[15]

Several studies have investigated the use of alfaxalone as a single dose, both for IV and IM routes. In terrapins and tortoises, alfaxalone at 5 mg/kg IV led to an anesthetic plan suitable for ETI and a surgical plane of anesthesia lasted for about 27 minutes.[16] In loggerhead sea turtles (*Caretta caretta*), three doses of alfaxalone (3 mg/kg, 5 mg/kg, and 10 mg/kg) were administered IV; all doses allowed ETI. At the highest doses all animals required positive pressure ventilation and recovery time increased with doses (28, 46, and 90 minutes, respectively). When alfaxalone was administered IV at 9 mg/kg in Macquarie River turtles (*Emydura macquarii)*, a suitable anesthetic plan for intubation was reached, but not a surgical plan of anesthesia. In addition, recovery time was shorter compared with other species (8-25 minutes).[17]

Gaining IV access in awake chelonian is not always possible. The main advantage of alfaxalone compared with propofol is that it can be administered IM. Studies investigating the use of alfaxalone IM have been performed in Horsfield's tortoises (*Agrionemys horsfieldii*)[18] and in red-eared sliders (*Trachemys scripta elegans*).[19] Doses of 10-20 mg/kg IM were used; the higher dose was needed when animals were maintained at higher temperatures, probably due to increased hepatic metabolism. When alfaxalone was given at 9 mg/kg IM to Macquarie River turtles, no sedation was obtained.[17]

### Clinical experience

The author has used alfaxalone as singe intramuscular and/or intravenous anesthetic agent on several occasions (eg, for sexing or for computed tomography in different species, including *Astrochelys radiata*, *Graptemys sp.* and *Caretta caretta*). If induction times are generally short, induction doses are higher and recovery times appear longer in turtles compared with tortoises.

For longer procedures, such as gastrotomy, ovariohysterectomy and laparoscopies, alfaxalone has been successfully used in TIVA leading to a stable anesthetic plan with shorter recovery times compared with procedures of similar length performed under inhalant anesthesia. Thus, it is the author's practice to use TIVA of alfaxalone when a long (> 25-30 minutes) anesthesia time is expected. The main difference noted by the author is a higher heart rate when alfaxalone is used compared with inhalant anesthesia. Further studies are warranted to elucidate differences in cardiovascular and respiratory parameters.

## References


1. Mosley CI, Mosley CA. Comparative anaesthesia and analgesia of reptiles, amphibians and fishes. In: Grimm KA, Lamont LA, Tranquilli WJ, Greene SA, Robertson S, eds. *Veterinary Anaesthesia and Analgesia.* Ames, IA: Wiley-Blackwell;2015:784-799.

2. Sladky K. Clinical anaesthesia in reptiles. *J Exot Pet Med.* 2012;21:17-31.

3. Giorgi M, Salvadori M, De Vito V et al. Pharmakokinetic/pharmacodynamics assessments of 10 mgkg tramadol intramuscular injection in yellow-bellied slider turtles (*Trachemys scripta scripta*). *J Vet Pharmacol Ther.* 2015;38:488-496.

4. James LE, Williams CJA, Bertelsen MF, Wang T. Anaesthetic induction with alfaxalone in the ball python (*Pyton regius*): dose response and effect of injection site. *Vet Anaesth Analg.* 2018;45:329-337.

5. Holz P, Barker IK, Burger JP, et al. The effect of the renal portal system on pharmacokinetic parameters in the red-eared slider (*Trachemys scripta elegans*). *J Zoo Wildl Med.* 1997;28:386-393.

6. Gans C, Gaunt AS. Visceral organs. In: Gans C, Gaunt AS, eds. *Biology of the Reptilia.* Vol 19. Ithaca, NY: Society for the Study of Amphibians and Reptiles;1998.

7. Glass M, Buggren WW, Johansen K. Ventilation in an aquatic and terrestrial chelonian reptile. *J Exp Biol.* 1978;72:165-179.

8. Schifino Valente AL, Martinez-Silvestre A, Garcia-Guasch L, et al. Evaluation of pulmonary function in European land tortoises using whole-body plethysmography. *Vet Rec.* 2012;171:154-158.

9. Bennet RA. A review of anesthesia and chemical restraint in reptiles. *J Zoo Wildl Med.* 1991;23:282-303.

10. Vigani A. Chelonia (Tortoises, Turtles and Terrapins). In: West G, Heard D, Caulkett N, eds. *Zoo Animal and Wildlife Immobilization and Anaesthesia.* 2nd ed. Ames, IA: Wiley-Blackwell; 2014;365-390.

11. Kastner SBR. Intravenous anaesthetics. In: Seymour and Duke-Novakovski, ed. BSAVA Manual of Canine and Feline Anaesthesia and Analgesia. 2nd ed. Gloucester, UK: British Small Animal Veterinary Association;2014;133-149.

12. Dewangan R, Tiwari SK. Total intravenous anaesthesia (TIVA) in veterinary practice. *Int J Sci Res.* 2015;1533-1539.

13. Suarez MA, Dzikiti BT, Stegmann FG, Hartman M. Comparison of alfaxalone and propofol administered as total intravenous anaesthesia for ovariohysterectomy in dogs. *Vet Anaesth Analg.* 2012;39:236-244.

14. Campagna I, Schwarz A, Keller S, et al. Comparison of the effects of propofol or alfaxalone for anaesthesia induction and maintenance on respiration in cats. *Vet Anaesth Analg.* 2015;42:484-492.

15. Vettorato E. Prolonged intravenous infusion of alfaxalone in a cat. *Vet Anaesth Analg.* 2011;40:551-556.

16. Knotek Z. Alfaxalone as an induction agent for anesthesia in terrapins and tortoises. *Vet Rec.* 2014;175:327-330.

17. Scheelings TF. Use of intravenous and intramuscular alfaxalone in Macquarie River turtles (*Emydura macquarii*). *J Herpetol Med Surg.* 2013;3-4:91-94.

18. Hansen LL, Bertelsen MF. Assessment of the effects of intramuscular administration of alfaxalone with and without medetomidine in Horsfield's tortoises (*Agrionemys horsfieldii*). *Vet Anaesth Analg.* 2013;40:e68-e75.

19. Shepard MK, Divers S, Braun C, Hofmeister EH. Pharmacodynamics of alfaxalone after single-dose intramuscular administration in red-eared sliders (*Trachemys scripta elegans*): a comparison of two different doses at two different ambient temperatures. *Vet Anaesth Analg.* 2013;40:590-598.

# Reptile Ophthalmology: Anatomy, Physiology, Diagnostics, and Therapeutics

Mikel Sabater, LV, RCVS CertZooMed, Specialist in Zoo and Wildlife, DipECZM (Avian), and Maria Luisa Pérez Orrico, LV, MRCVS, CES d'Ophtalmologie ENVT

**Session #2849**

*Affiliation:* From Exoticsvet, Valencia, Spain (Sabater), The Ralph Veterinary Referral Hospital, Marlow, UK (Orrico).

*Abstract*: Ocular conditions in reptiles are occasionally seen in the daily exotic animal practice. A complete ocular examination should be included in every physical examination as it may reveal a primary problem and/or the clinical manifestation of a systemic disease. The vast number of existing species results in a wide range of anatomic and physiologic adaptations that may be known when dealing with them. Ophthalmic disorders in reptiles can have an infectious or a non-infectious etiology; however, congenital abnormalities and disorders secondary to improper husbandry are overrepresented in this group of animals when compared to other exotic species. In addition, significant advances have been made recently in reptilian ophthalmology, especially in the field of diagnostics. This master class is structured in four parts. First, the description of the anatomic particularities of the eye and the adaptations presented in different groups of reptiles is described. Then, a comparison of the visual pathways and ocular physiology between different reptiles and other exotic species is used to explain the evolutionary adaptations observed in reptiles. After this, a review of the diagnostic procedures of potential utility to assess ocular conditions will be made before the medical and surgical treatment of ocular conditions is discussed. This master class is aimed to be useful for the exotic animal veterinarian with minimal information in ophthalmology as well as to those veterinarians with advanced skills in ophthalmology.

## Ocular Anatomy and Physiology

The ocular globes in reptiles are almost spherical and located lateral in the skull. In most species (chameleons would be an exception), their movements are very limited due to the poorly developed recti extraocular muscles. The degree of development of the ocular globe retractor muscle varies greatly between species. The scleral ossicles maintain the shape of the globe and participates in accommodation. Scleral ossicles are not observed in crocodilians which accommodate by changing antero-posteriorly the shape of the lens.

The upper eyelid is normally smaller and less mobile, except in crocodilians, than the lower, which has a fibrous tarsal plate in some lizard. The third eyelid is observed in most species and can also be transparent (more common in lizards but also in some chelonians (*Chelodina* sp. and *Emyda* sp.).

The spectacle is a nonmobile, transparent dermal structure formed by the fusion of the eyelids separated from the cornea by the subspectacular space. This is observed in snakes, Amphisbanidae, some geckos, Lacertilia, Teiidae, Uroplatus, Pygopodidae, and Xantusiidae. The spectacle has three layers (keratinized external epithelium, stroma and internal epithelium) and is highly vascularized. The spectacle opacifies during the majority of ecdysis and becomes transparent again approximately 24 hours prior to completing the shed. The subspectacular space is filled by the secretions of the Harderian gland and drains through a single duct to the oral cavity, close to the base of the vomero-nasal organ.

Lizards and chelonians have a well-developed Harderian gland in the orbito-nasal region that opens to the nictitans membrane. It produces the lipidic component of tears and has also an immune function. Salt glands are observed in some species but are absent in geckos and chameleons. The lacrimal gland produces the aqueous component of the tear and is in the orbito-temporal region. The nasolacrimal duct is present in most species (discrepancies exist for some *Testudo* sp.)

In the conjunctival flora of terrapins a predominance of gram-positive bacteria are observed, whereas gram-negative are more predominate in terrestrial ones.

The cornea of lizards is thin and arcuate and lacks the Bowman's membrane. The lens has a flat anterior capsule resulting in a deep anterior chamber. Some geckos lack a Descemet's membrane. Ophidians have thick and flat corneas with no Bowman's membrane, and thin anterior chambers. Chelonians have a thick cornea with a prominent Descemet's membrane.

The iris varies greatly between species and sexual dimorphism has been reported in some species (eg, box turtles). Nocturnal lizards have vertically-shaped pupils. Stenopeic pupillary openings are observed in geckos and crocodiles allowing a more defined image without accommodation when the pupil is miotic. Other pupillary shapes, such as the more or less horizontal pupils in some snakes (eg, East Indian long-nosed, tree snake or the African bird snake) have been reported.

The lens is surrounded by an annular pad. In lizards, they are soft and clear, except in some diurnal geckos. The annular pad of chelonians is less developed, the capsule is fine and the content almost fluid. Lizards and reptiles accommodate by deformation. The iridal muscles and the ciliary bodies are very-well developed in terrapins which combined with the soft lens allow a wide dioptric range by accommodating through an anterior-posterior displacement of the lens.

The retina is anangiotic (avascular) and receives its nutrition from choroidal vessels in chelonians or modified vascular structures protruding into the vitreous called the conus papillaris, in most lizards. Amphisbaenids lack a conus and, in crocodiles and some snakes, a regression of the conus to a preretinal vascular net known as the membrana vasculosa retinae, is observed. In others (eg, colubrids), the capillaries of this membrane penetrate the retina forming the intraretinal vessel. The amount of cones and rods varies between species depending of their visual needs. An area and a fovea centralis provide more visual acuity; however, those are not observed in crocodiles and is absent or reduced in nocturnal lizards. Crocodilians have a retinal tapetum formed by guanine crystals which are hyperreflective under dim light conditions.

An almost complete nerve decussation at the level of the optic chiasm has been reported in *Trachemys scripta elegans*.

Blindness in chelonians may result in anorexia due to their inability to locate food, varanids may compensate it almost completely with olfaction and snakes may do it with a thermoreception and the vomero-nasal organ.

## Ocular Examination

Always perform an ocular exam after a complete physical exam, including the oral cavity and ear canals.

Equipment includes a slit-lamp, tonometer, direct ophthalmoscopy, magnifying lenses (25-90D), fluorescein dye test, Schirmer test, among other things. Direct pupillary reflex is fast. Indirect pupillary reflex is minimal to absent, due to striated muscle. A Schirmer/red phenol thread test and paper point tear test (in species with eyelids) is indicated. A rebound tonometer is preferred over applanation in individuals with small corneal diameters.

Turtles should be held by the head and not by the neck to avoid impairing blood return from the head. Position of the ocular globe with respect to the body influences the intraocular pressure (IOP).

Mydriasis can be achieved under general anesthesia or after the intracamerular injection (27- to 30-gauge needle) of neuromuscular blockers (eg, D-tubocurarine, action 30 minutes to hours) after applying local anesthetic (eg, proxymethacaine) or under sedation. Topical vecuronium, especially when combined to phenylephrine, may be useful in chelonians. In some chelonians, topical proparacaine acts in one minute and last more than 45 minutes. Lidocaine infiltrations may be necessary in species able to close the eyelids strongly

Fundoscopy can be done with a panoptic or indirectly with a 90D lens. Other diagnostic tests include diagnostic imaging, microbiologic cultures, cytology and histopathology.

## Disorders of the Eyelids and the Conjunctiva

### Conjunctivitis

*Bacterial:* Primary rare. Caseous plaques may evolve to septicemia and death. *Chlamydia* sp., *Aeromonas* sp., *Pasteurella* sp., *Proteus* sp., *Citrobacter* sp., *Pseudomonas* sp., *Salmonella* sp., etc. May be predisposed by vitamin A deficiencies.

*Viral (HPV):* Fibropapilloma; lung, eye, tracheal disease (LETD) in green sea turtles.

*Parasitic:* Leeches, *Neopolystoma* spp, *Laraedius laraedii.*

*Post-hibernation:* Freezing.

*Others:* Contact chemical irritation (eg, some plants), trauma, foreign bodies.

### Blepharedema and blepharitis

*Traumatic:* Causes can include trauma or post-traumatic injury.

*Nutritional:* Vitamin A deficiency. Squamous metaplasia of the epitheliums (ocular orbit glands or ducts, renal, pancreatic, respiratory and gastrointestinal). Frequently seen in chelonians and in *Anolis carolinensis*. The squamous metaplasia predisposes animals to secondary bacterial infections. Correct nutritional deficiencies, supplement with parenteral vitamin A (avoid overdosing) if needed, and provide antibiotics if their use is justified.

*Bacterial:* Infections normally manifest as blepharospasm and skin discoloration.

- Primary: Mycoplasmosis. *Gopherus polyphemus* (*Mycoplasma agassizii*) and American alligators (*M. alligatoris*). Clarithromycin.
- Secondary: Abscesses (gram-negative [*E. coli, Pseudomonas* spp., *Aeromonas* spp.] and acid-fast). Septicemic cutaneous ulcerative disease (SCUD). *Citrobacter freundii* in terrapins.

*Parasitic: Foleyella* spp., *Trombicula* sp., and *Hirstiella* sp. In snakes, the mite *Ophionyssus natricis*, produces pruritus.

*Fungal: Penicillium* sp. Plaques. Antifungals, chlorhexine. Candidiasis are opportunistic.

*Viral:* Herpesvirus in *Chenoloidis chilensis* and ref-footed tortoises.

Diagnostic: Cytology, microbiologic cultures and sensitivities, PCR (HPV, *Mycoplasma* sp.).

Treatment: Disinfection with iodine povidone (non-alcoholic), topical and sometimes also systemic antibiotics (eg, tetracyclines, ciprofloxacine and gentamycin), surgical resection of granulomas.

*Neoplasia:* Viral (papillomas) and non-viral.

Diagnostic: Histopathology.

*Environmental:* Opacities of the third eyelid of adult alligators due to the accumulation of salts secondary to poor hygienic environments in captivity.

## Disorders of the Speculum

### Abrasions and ulcers

Opacities, neovascularization.

- Small abrasions may be seen during ecdysis.
- Trauma: aggressions, chemical irritants, bacterial and parasites.
- Fungal (*Fusarium* sp. in a *Boa arcoiris*). Chlorhexidine and ketoconazol.

Diagnostic: Cytology, microbiologic cultures and biopsies.

Treatment: Prevent corneal desiccation, topical antibiotics, N-acetyl-cysteine...

### Spectacle loss

Trauma: Iatrogenic, aggression.

Partial or total: Favor corneal desiccation which may end in panophthalmitis.

Treatment: Artificial tears, antibiotics, contact lenses, buccal mucous transposition.

### Infections

*Bacterial:* Gram-negative bacteria (more frequent).

*Fungal:* Topical creams. Single case of penetrating infection with *Fusarium oxysporum.*

*Sub-spectacular abscess:* Secondary to trauma, ascending infections through the nasolacrimal duct or hematogenous spread. Leukocoria (turbid to whitish spectacle). Microbiology. Corneal wedge resection to flush the subspectacular space with antibiotics (cipro or genta) and/or antifungals.

*Parasitic:* Mites (*Ophionissus*) and ticks (*Ixodidae*). Treatment: ivermectin (potentially toxic in some chelonians and snakes), fipronil, selamectine, etc. Nematodes (sub-spectacular (*Serpentirhabdias* spp. in *Python regius*). Protozoa (metronidazole).

*Retained spectacle:* In snakes, frequently associated with disecdysis due to incorrect environmental conditions (humidity, temperature, hygienic, malnutrition. In geckos, the most commonly diagnosed cause is bacterial infection. Slit-lamp. Treatment: Hydration of the retained spectacle, artificial tears, correct environmental humidity, provide baths, apply N-acetyl-cysteine. Do not force the removal of the retained spectacle.

*Pseudobuphthalmos:* Impaired lacrimal drainage in those species having a spectacle provokes the accumulation of tears in the subspectacular space (also referred in literature as precorneal dacryocele). Diagnostic: Injection of fluorescein in the lateral canthus of the spectacle avoiding excessive pressure over the cornea. Congenital or acquired (respiratory, oral, nasal). Temporal or permanent. Flushing or conjuntivoralostomy.

## Disorders of the Nasolacrimal Duct

Obstructions and dacryocystitis secondary to stomatitis, periorbital masses, hipovitaminosis A, and helminthiasis, among other causes.

### Dacryops

Unilateral exophthalmos, episcleral hyperemia, periocular inflammation, incomplete closure of the eyelids, and reduced ocular motility. Treatment is surgical.

### Keratoconjunctivitis sicca

Harderian gland (vitamin A deficiency).

Treatment: Correct the cause if known, and use artificial tears.

## Disorders of the Cornea

### Corneal edema

The cause can be infectious or noninfectious (eg, traumatic). It also occurs in marine species kept in fresh water. Treatment is to correct the cause and consider the use of antiedema ophthalmic drops.

### Keratitis

The causes are described below.

- *Testudo* spp. (*Moraxella, Pseudomonas, Aeromonas*). Surgical resection. Microbiology. Antibiotics (eg, ciprofloxacin) long term.
- Snakes: mycotic keratitis reported. Treatment varies depending on the cause.
- Viral infections
- LETD provokes corneal opacities.
- URTD (upper respiratory tract disease), HPV in terrestrial tortoises associated or not with mycoplasmosis. Ocular discharge, stomatitis and corneal ulcers. Topical acyclovir effectivity not demonstrated.

### Lipid keratopathy

Sometimes hereditary predisposition.

### Post-hibernation keratopathy

Protein granules in the cornea. NAC.

### Ulcers

Causes include foreign bodies, trauma (UVB), infections, etc.

Diagnosis: Fluorescein dye test.

Treatment: Topical antibiotics TID. Surgery if needed ( 8/0-10/0 sutures, surgical microscope) or blepharorhaphy.

## Disorders of the Uvea

### Uveitis

The cause can be infectious or noninfectious (eg, traumatic).

Treatment: Correct the cause and NSAIDs.

### Hyphema and hypopyon

Secondary to systemic infections (*Klebsiella pneumoniae* or *Pseudomonas* spp.).

Treatment: Antibiotics (eg, ciprofloxacin) and anti-inflammatories (eg, flurbiprofen and diclofenac). Enucleation.

### Anisocoria

This is potentially suggestive of inclusion body disease (IBD) in boids, retrovirus. Iridal cysts.

### Cataracts

The cause can be seniles, secondary to uveitis, UV. In *Testudos* sp., excessively low temperatures during hibernation may result in cataracts, vitreal opacities and CNS deficits.

Diagnosis: Ophthalmoscopy, slit-lamp examination, ultrasonography.

Treatment: Sometimes reversible. Also described in terrapins. Phacoemulsification reported in *Varanus exanthematicus*.

## Retinal Disorders

Probably underdiagnosed. Retinal degeneration secondary to vitamin A deficiency, thrombosis, and septicemia. Mercury intoxication.

## Disorders of the Ocular Globe

### Congenital abnormalities

Anophthalmia, microphthalmia, cystic eyes, cyclopia, colobomas, ankyloblepharon, afakia, atresia of the lacrimal duct, dermatization of the spectacle, color mutations.

### Exophthalmos

Retrobulbar masses: hematomas, cellulitis, abscesses or neoplasia. Cytology, diagnostic imaging, microbiology.

Associated with metabolic bone disease/ secondary nutritional hyperthyroidism.

### Thiamine deficiency

Affects the innervation and may produce a retraction of the ocular globe impairing its movement. More common in herbivore reptiles fed in vegetables rich in thiaminases or other species fed on frozen fish. Supplement with thiamine and correct the diet.

### Infectious

Bacterial or fungal. May result in phthisis bulbi.

### Enucleation

Slow healing (minimum 3 weeks). In snakes and geckos, it may be useful to dissect the scales from the periocular bone to obtain enough tissue to close the gap left after the eye has been removed.

# Ophidian Serpentoviruses (Nidoviruses): What We Know, What We Don't Know, and What We are Figuring Out

Robert J. Ossiboff, DVM, PhD, DACVP

**Session #2880**

*Affiliation:* From the Department of Comparative, Diagnostic, and Population Medicine, College of Veterinary Medicine, University of Florida, 2015 SW 16th Ave, Gainesville, FL 32610, USA.

*Abstract:* Ophidian serpentoviruses, previously referred to as nidoviruses, are increasingly being identified as important agents of clinical disease in captive snake populations. Serpentovirus infections can cause severe and sometimes fatal upper respiratory and upper alimentary disease that can result in significant losses to private and zoological collections. The objective of this presentation is to introduce the audience to the history of the viruses in snakes (and other reptiles), and detail comparative clinical, gross, and histopathologic findings in various snake species naturally infected with ophidian nidoviruses from both published and unpublished data. The current diversity of serpentoviruses known of reptiles will be presented, including which species are more commonly found to be clinically affected, and what that means for large and mixed snake collections. A review of diagnostic options and developing tests and future research directions will also be presented. The intent of this talk is to present the information known about these viruses to clinicians in a way that focuses on features of clinical and management significance.

## Introduction

Since the first reports of serpentovirus infections in snakes (previously referred to as nidoviruses) in 2014,[1-3] much has been learned about these important pathogens. Serpentoviruses are most commonly associated with disease in pythons,[1-5] but can also infect other species of boids.[6] Rarely, they can also be detected by molecular screening in colubrids (R.O., oral communication, April 2019). They are known to infect captive snakes in North America[2-3] and Europe,[1,4,6-7] though both wild reservoirs of related viruses in native and feral, free-ranging boid species are suspected. Serpentoviruses can be an important cause of morbidity and mortality in captive collections, and as such represent an important emerging snake pathogen that reptile veterinarians should be familiar with. An overview of the clinical presentation, postmortem findings, diagnostic testing options, and biosecurity recommendations will follow in this brief review of ophidian serpentovirus infections.

## Clinical Signs of Serpentovirus Infection

Serpentovirus infections manifest as oral and upper respiratory disease in pythons. Early signs of infection include reddening of the oral mucosa,[5] excess oral mucus production,[5] and may be associated with audible clicks and wheezes on respiration. The production of oral and nasal mucus can be copious, and may be evident on the sides of enclosures. The presence of excessive mucus is an important diagnostic feature, as it is something not commonly associated with other common viral respiratory infections in snakes, such as paramyxoviruses and reoviruses. Advanced serpentovirus infection is commonly associated with severe stomatitis and pneumonia with secondary Gram-negative bacterial infections.[2-4] While antimicrobial therapy may be palliative in the short term, signs often recur. Evidence of severe respiratory distress, including abnormal posturing, open-mouth breathing, and continued excessive mucus production, are commonly observed in advanced cases (R.O., oral communication, April 2019).

## Postmortem Findings

On gross necropsy, serpentovirus lesions are present in the oral and nasal cavities, the trachea, and, variably the orad (proximal) esophagus, and lungs.[2-5] Necrotizing stomatitis and nasopharyngitis (choanal) with associated plaques of necrotic and exudative material are often the result of concurrent viral and bacterial infections.[2-3] Abundant mucus or mucoid exudate is often present in the oral cavity, choanal, glottis/trachea, and esophagus. The lungs are often thick, edematous, and dark red, and may have the accumulation of mucus and/or exudate within the central lumen or faveoli that can extend back to the membranous portion of the caudal lung. Histologically, serpentovirus infections are characterized by proliferative and necrotizing information of the oral, nasal, pharyngeal, proximal esophageal, sinus, and tracheal mucosa as well as proliferative pneumonia with prominent lymphocytic inflammation.[2-5] Complicating secondary bacterial infections will add components of granulocytic inflammation and exudate accumulation in the aforementioned lesions.[2,3] The documentation of proliferative inflammation of the proximal esophagus is an important diagnostic feature of serpentovirus infections not seen in paramyxovirus or reovirus infections. Viral inclusion bodies are not a reported feature of serpentovirus infections, and while viral inclusion-like structures can sometimes be seen in some cases, they are not a reliable diagnostic feature (R.O., oral communication, April 2019). As lung involvement may not be striking in early serpentovirus cases (particularly if animals are sacrificed for diagnostic screening), histologic examination of the glottis, trachea, oral mucosa, tongue, and proximal esophagus in addition to the lung are all recommended for serpentovirus diagnosis, and can easily be examined in a single sagittal section of the ventral oral cavity.

## Diagnostic Testing Options

Antemortem diagnosis of serpentovirus infection is readily attainable of degenerate reverse-transcription polymerase chain reaction (RT-PCR testing) of a variety of samples. Swabs of the choanal and proximal esophagus using a polyester/rayon tipped, plastic shaft swab are preferred. Other samples, such as general swabs of the oral cavity, and tracheal washes are also good samples, though bacterial contamination (oral swabs) and dilution effects (tracheal washes) may decrease assay sensitivity (R.O., oral communication, April 2019). For postmortem sampling, frozen lung, trachea, and proximal esophagus are recommended. A number of private diagnostic and academic diagnostic laboratories offer RT-PCR testing. The most important thing to be aware of is the diversity of serpentoviruses identified to date. Labs offering specific RT-PCR or quantitative (real time) PCR tests may be able to detect specific viruses (such as ball python or green tree python serpentoviruses), but may miss other viruses. Since single hosts (such as ball pythons and green tree pythons) can be infected with a variety of serpentoviruses, broadly reactive degenerate RT-PCR is the preferred method for initial diagnostics. If candidate viruses are identified, then collection screening and surveillance by real time PCR becomes more useful.

## Biosecurity Recommendations

One of the most commonly encountered questions regarding serpentovirus infections in snake collections is what measures can be taken to control the spread of the virus, which has strong potential for causing high morbidity and mortality within collections.[3] Unfortunately, much of the background experimentation needed to accurately answer this question has not yet been performed; however, with the successful isolation and culture of this virus on susceptible cell lines,[3] active research projects to establish assays to determine viral infectious titers and how those titers change in response to environmental conditions and susceptibility to common disinfectants are ongoing. In situ hybridization (ISH) visualization of viral nucleic acids in diseased snakes with multiple ophidian serpentoviruses documents the presence of abundant virus within the oral and respiratory mucoid secretions

(R.O., oral communication, April 2019), suggesting a strong potential for droplet transmission. Moreover, swallowed mucoid discharge and the detection of virus within the lumen of the gastrointestinal tract also suggests the potential for fecal viral transmission (R.O., oral communication, April 2019). As such, collection owners should be advised of the broad susceptibility of pythons (and some boids) to these viruses, and the potential for spread between different host species. Practices to minimize movement of material between enclosures/tubs (especially uneaten food items) and proper disinfection protocols should be encouraged. Antemortem screening and isolation of RT-PCR positive animals may also mitigate the effects of these pathogens in large collections. Appropriate quarantine protocols for new acquisitions with successive rounds of antemortem diagnostic screening are recommended before introduction of new animals into established groups.

## Summary

Ophidian serpentoviruses (nidoviruses) are an important group of respiratory pathogens for which there is rapid evolution of veterinary medicine's understanding of their significance to captive snake collections. While there are limited treatment options for affected snakes, the diagnostic features of serpentovirus infections are distinguishable from other snake viral pathogens. With proper biosecurity and quarantine protocols, veterinarians and hobbyists/owners can work together to limit the potential effect of these viruses on captive animals.

## References


1. Bodewes R, Lempp C, Schürch AC, et al. Novel divergent nidovirus in a python with pneumonia. *J Gen Virol.* 2014;95:2480-2485.

2. Stenglein MD, Jacobson ER, Wozniak EJ, et al. Ball python nidovirus: a candidate etiologic agent for severe respiratory disease in *Python regius*. *mBio.* 2014;5:e01484-1414.

3. Uccellini L, Ossiboff RJ, de Matos, REC, et al. Identification of a novel nidovirus in an outbreak of fatal respiratory disease in ball pythons (*Python regius*). *Virol J.* 2014;11:144.

4. Dervas E, Hepojoki J, Laimbacher A, et al. Nidovirus-associated proliferative pneumonia in the green tree python (*Morelia viridis*). *J Virol.* 2017;91:e00718-17.

5. Hoon-Hanks LL, Layton ML, Ossiboff RJ, et al. Respiratory disease in ball pythons (*Python regius*) experimentally infected with ball python nidovirus. *Virology.* 2018;517:77-87.

6. Marschang RE, Kolesnik E. Detection of nidoviruses in live pythons and boas. *Tierarztl Prax Ausg K Kleintiere Heimtiere.* 2017;45:22-26.

7. Rampacci E, Masi M, Origgi FC, et al. First molecular detection of ball python nidovirus in Italy. *Acta Veterinaria Hungarica.* 2019;67:127-134.

# Reptile Emergency and Critical Care: Current Updates from the Literature

Olivia A. Petritz, DVM, Dipl ACZM, and
Krista A. Keller, DVM, Dipl ACZM

**Session #2633**

*Affiliation:* From North Carolina State University, College of Veterinary Medicine, 1060 William Moore Drive, Raleigh, NC 27607, USA (Petritz), University of Illinois College of Veterinary Medicine, 1008 W Hazelwood Drive, Urbana, IL 61802, USA (Keller).

## Introduction

Reptile medicine is a growing field with advances made in each issue of each journal we read. We present some of the recent non-anesthetic advances in reptile medicine that have changed the way that our industry has approached treatment in this diverse group of animals.

## Oxygen Therapy

In mammals, hypercapnia stimulates respiration; however, in reptiles, respiration is controlled by hypoxemia, hypercapnia, and environmental temperature.[1] Many species differences exist; however, in most species, low oxygen concentrations increase respiration rate and hypercapnia increases tidal volume. Several studies have evaluated the clinical ramifications of these differences including an unpublished study in green iguanas (*Iguana iguana*), where prolonged recovery from anesthesia was noted in patients that received 100% oxygen versus room air.[2] The findings from this study set the stage for the mantra that reptiles should rarely be administered oxygen therapy and should be recovered from anesthesia in room air. More recent studies in bearded dragons (*Pogona vitticeps*)[3,4] and Dumeril's monitors (*Varanus dumerili*),[5] found minimal if no statistical difference between animals recovered in room air (approximately 21% oxygen) and 100% oxygen. An additional study has shown that administration of 100% inspired oxygen resulted in a significant increase in $PaO_2$ in conscious green iguanas, with no inhibitory changes in respiratory rate.[6] Although older literature supports that reptiles administered oxygen have slower anesthetic recoveries, the more recent studies demonstrate that oxygen administration is not harmful to reptilian patients, does not slow anesthetic recovery in the species examined, and should be considered for patients exhibiting respiratory compromise, as in mammals and birds.

## Advances in Fluid Therapy

### Routes of administration

Whereas in other species, route of administration for fluid therapy is decided based on level of dehydration, choices for route of administration in many reptile patients is limited to enteral, cloacal and subcutaneous routes due to the historic challenges in anatomy to provide more aggressive fluid therapy. Recent literature has elucidated techniques to provide an intravenous (IV) and intraosseous (IO) route of administration in chelonian patients.

A study in desert tortoises (*Gopherus agassizii*) comparing IO injections in several sites has shown that humeral and femoral IO catheter administration was able to achieve high bioavailability (84% and 62%, respectively) of fluids. IO catheters placed in the bridge or gular site of the shell had much lower bioavailability of fluids, (42% and 41%, respectively). This study supports that IO catheters placed into the humerus or femur are no more technically challenging than IV catheters and are a viable option for aggressive fluid therapy.[7]

A recent study reports on a novel technique of providing long-term venous access in chelonian patients. The case series reports on 3 large species of chelonians (7.7-24.6 kg) and the procedure is performed under sedation or anesthesia. A sterile cut down procedure allows access to the jugular vein then, catheterization with a standard 20-gauge over-the-needle catheter is performed that allows introduction of a guide wire for a 4-Fr 13-cm double-lumen central venous catheter. The wire stays in place and the introducing catheter is removed, then a vein dilator is threaded and introduced into the vein and lastly the jugular catheter is threaded through the guide wire to a predetermined length. After the guide wire is removed, the catheter is sutured in place and covered in transparent dressing. The chelonians reported had these catheters in place for many reasons for time periods spanning from 1-4 months with no complications reported. Benefits of this technique include the ability to provide fluid therapy or drug administration directly into the venous system, measure central venous pressure and to have repeated access to the venous system for hematologic samples.[8]

### Choosing appropriate fluids

In mammals, 60% of an animal's body weight is water while 40% of the body weight is within the intracellular fluid compartment and 20% is within the extracellular fluid compartment.[9] In non-chelonian reptiles, 71-75% of the body weight is fluid, with an equal distribution between the intracellular and extracellular fluid compartments.[10] The carapace and plastron of chelonians reduce their distribution of total body water closer to that of mammals, approximately 66%. An increase in extracellular fluid plasma osmolality will result in fluid shifting out of cells causing cellular dehydration. A decrease in extracellular plasma osmolality will cause fluid to shift into cells, resulting in cellular swelling. Rapid fluid shifts in either direction can have detrimental consequences.[9,10]

Osmolality refers to the number of osmoles per kilogram of solvent (expressed as milliosmoles per kilogram, mOm/kg), and osmolarity refers to the number of osmoles per liter of solution (expressed as milliosmoles per liter, mOm/L).[11] In biologic fluids, the difference in osmolality and osmolarity is negligible, and the terms often are used interchangeably. Knowledge of a reptile species normal osmolarity as well as a patient's osmolarity on emergency presentation would be helpful in determining the proper fluid therapy and minimizing detrimental fluid shifts. Most of the intravenous fluid products readily available on the market are formulated based on the plasma osmolarity of humans. Selection of the appropriate crystalloid solution for reptiles is confounded by a paucity of information about species-specific plasma osmolarity, and studies supporting species-specific osmolality are presented in Table 1. These differences should be considered when administration of large volumes of fluids are needed to avoid rapid and large shifts in the patient's plasma osmolarity. Ideally the use of crystalloid solutions that are significantly hypo- or hyper-osmolar to expected osmolarity should be avoided.

**Table 1.** Studies reporting species-specific osmolality in reptilian patients.

| Species | Reported Osmolality (mOsm/L) | Reference |
|---|---|---|
| Bearded dragon (*Pogona vitticeps*) | 295.4 ± 9.35 | 12 |
| Green iguana (*Iguana iguana*) | 327 ± 3.3 | 13 |
| Gila monster (*Heloderma suspectum*) | 284-298 | 14 |
| American alligator (*Alligator mississippiensis*) | 269.3 | 15 |
| Corn snakes (*Pantherophis guttata*) | 304.5-373.0 | 16 |

Determination of plasma osmolality can be performed either by direct measurement with an osmometer or by calculation based on formulas utilizing biochemistry values. Studies have shown that directly measured plasma osmolality in reptiles had poor agreement with calculated values.[12,15] The authors of these studies do not recommend calculating plasma osmolality using previously reported formulas for mammals.

## Whole Blood Transfusion

Whole blood transfusions are recommended in case of acute severe hemorrhage and life-threatening anemia, regardless of species and a recent study has identified a novel method to evaluate the presence of anemia in Eastern box turtles (*Terrapenne carolina carolina*) using the iris color.[17] Ideally a whole blood donor of the same species would be utilized; however, given the many thousands of different reptile species, finding a conspecific donor may not always be possible. Heterologous blood transfusions from a reptile of a different species may still be beneficial and are associated with minimal complications. Few studies report on the changes of stored whole blood in reptiles. In American alligators (*Alligator mississippiensis*), a significant decrease in PCV from day 7 to 35, as well as increases in osmotic fragility after 45 days of storage are reported.[18] In loggerhead sea turtles (*Caretta caretta*), whole blood stored in CPDA-1 and sodium citrate had more severe electrolyte and pH changes than those stored in sodium heparin.[19] These studies outline that species differences exist with regards to changes during storage of whole blood and that fresh whole blood is the most ideal donation to be administered. These guidelines, paired with the presented information regarding IV and IO access will facilitate blood transfusions in reptilian species.[7,8]

Incompatibility reactions including red blood cell agglutination between both homologous and heterologous donor-recipient pairs have been reported, therefore it would deem most advisable to perform crossmatches whenever possible.[20] Up to 10% of the total blood volume can safely be removed from the reptile donor and a donor should have baseline CBC performed to ensure it is safe to remove that blood volume prior to donation. Several case reports have described the successful use of homologous and heterologous blood transfusions in several reptile species[20-25] Complications secondary to blood transfusions have been infrequently reported including orange discoloration to the urine in a Hispaniolan slider (*Trachemys decorata*)[21] and increased WBC was noted after two rapid blood transfusions from a different species; this leukocyte response was suspected to reflect a possible destructive immune reaction, although no clinical recognizable signs of reactions were observed.[24]

In dogs, cats, and humans, the use of a blood filter such as the Hemo-Nate 18-µm filter (Utah Medical Products, Inc, Midvale, UT, USA) is the standard of care during a blood transfusion. In a study of the effects of an 18-µm Hemo-Nate filter during whole blood transfusion in American alligators, no hemolysis or significant decreases in the packed cell volume were noted secondary to the use of the filter, despite the larger size of reptilian erythrocytes compared to mammalian.[26] The authors of the study recommended the use of a Hemo-Nate filter during administration of whole blood in reptiles.

## Antibiotic Stewardship

A recent conference proceeding has opened the discussion for developing institution- and/or industry-specific statements on antibiotic stewardship. This discussion spans all taxa, but is brought forth here to further the discussion in reptilian patients. The presentation discussed the institutional policy developed at UGA-VTH where antibiotics are only prescribed to patients after proving an infectious process is ongoing. Minimal diagnostics are required prior to prescription of any antibiotic, and must include cytologic proof of septic inflammation.

Additional diagnostics that may prove an infectious process may include histopathology or culture and sensitivity. The policy also provides guidelines for judicious use of antibiotics whether prescribed empirically or based upon culture and sensitivity results. For instance, third-generation cephalosporins should be reserved for cases in which culture and sensitivity results have proven resistance to first and second generation cephalosporins. In a time where antibiotic resistance is an ever-present issue in human and veterinary medicine, all facets of veterinary medicine have a responsibility to responsibly use this class of medications.[27]

## References


1. Schumacher J. Respiratory medicine of reptiles. *Vet Clin North Am Exot Anim Pract.* 2011;14:207-224.

2. Diethelm G, Mader D. The effects of $FIO_2$ on post anesthetic recovery times in the green iguana. *Proc Annu Conf Assoc Reptilian Amphib Vet.* 1999;169-170.

3. Odette O, Churgin SM, Sladky KK, Smith LJ. Anesthetic induction and recovery parameters in bearded dragons (*Pogona vitticeps*): comparison of isoflurane delivered in 100% oxygen versus 21% oxygen. *J Zoo Wildl Med.* 2015;46(3):534-540.

4. Ratliff C, Parkinson LAB, Mans C. Effects of the fraction of inspired oxygen on alfaxalone-sedated inland bearded dragons (*Pogona vitticeps*). *Am J Vet Res.* 2019;80(2):129-134.

5. MF Bertelsen, Mosley C, Crawshaw GJ, et al. Inhalation anesthesia in Dumeril's monitor (*Varanus dumerili*) with isoflurane, sevoflurane, and nitrous oxide: effects of inspired gases on induction and recovery. *J Zoo Wildl Med.* 2005;36:62-68.

6. Hernandez SM, Schumacher J, Lewis SJ, et al. Selected cardiopulmonary values and baroreceptor reflex in conscious green iguanas (*Iguana iguana*). *Am J Vet Res.* 2011;72(11):1519-1526.

7. Young BD, Stegeman N, Norby B, Heatley JJ. Comparison of intraosseous and peripheral venous fluid dynamics in the desert tortoise (*Gopherus agassizii*). *J Zoo Wildl Med.* 2012;43(1):59-66.

8. Pardo MA, Divers S. Jugular central venous catheter placement through a modified Seldinger technique for long-term venous access in chelonians. *J Zoo Wildl Med* 2016;47(1):286-290.

9. Wellman M, DiBartola S, Kohn C. Applied physiology of body fluids in dogs and cats. In: DiBartola S, ed. *Fluid, Electrolyte, and Acid-Base Disorders in Small Animal Practice.* St. Louis, MO: Elsevier Saunders;2012:2-4.

10. Mader D, Rudloff E. Emergency and critical care. In: Mader D, ed. *Reptile Medicine and Surgery.* 2nd ed. St Louis, MO: Elsevier Saunders;2006:533-548.

11. DiBartola S. Applied physiology of body fluids in dogs and cats. In: DiBartola S, ed. *Fluid, Electrolyte, and Acid-Base Disorders in Small Animal Practice.* St. Louis, MO: Elsevier Saunders;2012:8-10.

12. Dallwig RK, Mitchell MA, Acierno MJ. Determination of plasma osmolality and agreement between measured and calculated values in healthy adult bearded dragons (*Pogona vitticeps*). *J Herpetol Med Surg.* 2010;20:69-73.

13. Maxwell L, Helmick K. Drug dosages and chemotherapeutics. In: Jacobson E, ed. *Biology, Husbandry, and Medicine of the Green Iguana.* Malabar, FL: Krieger Publishing;2003:131-151.

14. Davis JR, DeNardo DF. The urinary bladder as a physiological reservoir that moderates dehydration a large desert lizard, the Gila monster, *Helodema suspectum*. *J Exp Biol.* 2007;210:1472-1480.

15. Nevarez JG, Acierno MJ, Angel M, et al. Determination of agreement between measured and calculated plasma osmolality values in captive-reared American alligators (*Alligator mississippiensis*). *J Herpetol Med Surg.* 2012;22:36-41.

16. Guzman DS, Mitchell MA, Acierno M. Determination of plasma osmolality and agreement between measured and calculated values in captive male corn snakes (*Pantherophis guttatus guttatus*). *J Herpetol Med Surg.* 2011;21(1):16-19.

17. Cerreta AJ, Mehalick ML, Stoskopf MK, et al. Assessment of a visual scoring system for identifying and quantifying anemia in male eastern box turtles (*Terrapene carolina carolina*). *J Zoo Wildl Med.* 2018;49(4):977-982.

18. Emerson, JA, Stacy NI, Coverdill CC, et al. Preservation of red blood cells in the American alligator (*Alligator mississippiensis*) over time and in two different anticoagulants. *J Herpetol Med Surg.* 2014;24(3):82-94.

19. Phillips BE, Stoskopf M, Beasley JF, Harms CA. Evaluation of three anticoagulants use for short term storage of loggerhead sea turtle (*Caretta caretta*) whole blood. *J Herpetol Med Surg.* 2017;27(3-4):97-103.

20. Wu C, Chie C. Case study: successful treatment of acute anaemia with whole-blood transfusion and supportive care in two turtles. *Proc Exot Avian Conf: Topical Tropical*;2014.

21. Wack R, Anderson N. Resuscitation of a Hispaniolan slider, *Trachemys decorata*, using oxyglobin and a blood transfusion. *J Herpetol Med Surg.* 2004;4.

22. Zachariah T, Langan J, Kinsel M. Presumed hypovitaminosis A secondary to chronic bile duct obstruction in a Hermann's tortoise (*Testudo hermanni*). *Proc Annu Conf Assoc Reptilian Amphib Vet.* 2010;86.

23. Gibbons P, Darbo-McLeffan H. Homologous blood transfusion in an African spurred tortoise (*Geochelone sulcata*). *Proc Annu Conf Assoc Reptilian Amphib Vet.* 2009;138-139.

24. McCracken H, Hyatt A, Slocombe R. Two cases of anemia in reptiles treated with blood transfusion (1) hemolytic anemia in a diamond python caused by an erythrocytic virus, (2) nutritional anemia in a bearded dragon. *Proc Annu Conf Assoc Reptilian Amphib Vet.* 1994;47-51.

25. Mehalick M, Lewbart G, Christian L. Multiple blood transfusions in an eastern box turtle. *Proc Annu Conf Assoc Reptilian Amphib Vet.* 2007;57-58.

26. Nevarez, Cockburn J, Kearney MT, Mayer J. Evaluation of an 18-micron filter for use in reptile blood transfusions using blood from American alligators (*Alligator mississippiensis*). *J Zoo Wildl Med.* 2011;42(2): 236-240.

27. Divers S, Sladakovic I, Mayer J, Sanchez S. Development of an antibiotic policy in a zoological medicine service and approach to antibiotic dosing using minimum inhibitory concentration data. *Proc Assoc Exot Mammal Vet, Assoc Reptilian Amphib Vet.* 2017;36-44.

**Gap in pagination due to formatting issues.**

**Page 619**

# Reptilian and Amphibian Reproductive Cycles, Eggs and Embryos, Normal Development and Incubation

Jeanette Wyneken, PhD

**Session #2744**

*Affiliation:* From the Department of Biological Sciences, Florida Atlantic University, 777 Glades Rd, Boca Raton, FL 33431, USA.

The development of reptiles and amphibians has been studied in a fraction of species, and is most thoroughly described for those that can be propagated in captivity as well as some protected species. There have substantial declines in amphibians world-wide and major declines in many reptile species. Amphibians are particularly vulnerable climatic conditions due to their moist and delicate skin. Amphibians might be highly susceptible to captive climates presented and to climate change, in spite of their evolutionary history as long-term survivors. Many have very narrow ranges with unique combinations of habitat requirements.[1] Consequently, knowledge of normal development and developmental processes is of fundamental importance for both captive and wild amphibians and reptiles.

In both amphibians and reptiles, there are two major characteristics that classify embryonic development. The first is where the embryo develops and second, how it is nourished.[2] The major traits can affect the reproductive health of the parent(s) and the success of reproduction. Where the embryo develops is described by ovipary and vivipary. Ovipary refers to egg-laying reproduction. It occurs in all reptilian families: Testudinata (turtles and tortoises), Lepidosauria (Squamata; lizards and snakes), Crocodilia (alligators, caimans, crocodiles, and garials) and Sphenodontia (tuatara). Vivipary refers to live-bearing reproduction and is restricted to the Lepidosauria. The term ovo-vivipary is antiquated and not appropriate because it confounds where the embryo develops with the extent to which it gets its nutrients from the yolk or from the mother across the vitelline membranes.

Blackburn (2015) reports that vivipary (live-bearing reproduction) has arisen independently in more than 150 vertebrate lineages, including a minimum of 115 clades of extant squamates.[3] Approximately 20% of the lizards and snakes are viviparous. There are no viviparous chelonians, crocodilians, or tuataras.

In oviparous development, the embryo receives nutrients via lecithotrophy, in which yolk from the ovum provides nutrients for development. In viviparous species, embryonic nutrition is matrotrophy, such that maternal provisioning occurs during gestation in the reproductive tract.[2] Matrotrophy arose independently in vertebrates at least 33 times so there can be considerable diversity. Matrotrophy has multiple independent origins and diverse forms, many of which are not fully distinct because embryos can transition from one form to another. Matrotrophy, includes placentotrophy (in which nutrition and gas exchange cross a placenta formed by fetal membranes in contact with uterine wall) and is found in a number of skink species (Scindae); for example, several species of *Chalcides, Eumecia*, and *Mabuya*.[3] Histotrophy is a form of matrotrophy (in which embryos ingest nutrients, typically liquids of maternal origin, that diffuse across tissue spaces to absorptive sites of the embryo); oophagy (in which embryos develop in the reproductive tract and are nourished by ingesting sibling yolks or yolk breakdown products); matrophagy (in which embryos actively eat maternal tissues such as the oviductal mucosa or sections); and embryophagy (developing embryos obtain nutrition by cannibalizing siblings.[3]

Here I describe basic patterns of reproduction, eggs, and embryonic development. There are substantial differences between amphibians and reptiles (Fig 1). Consequently much is discussed separately for the two clades.

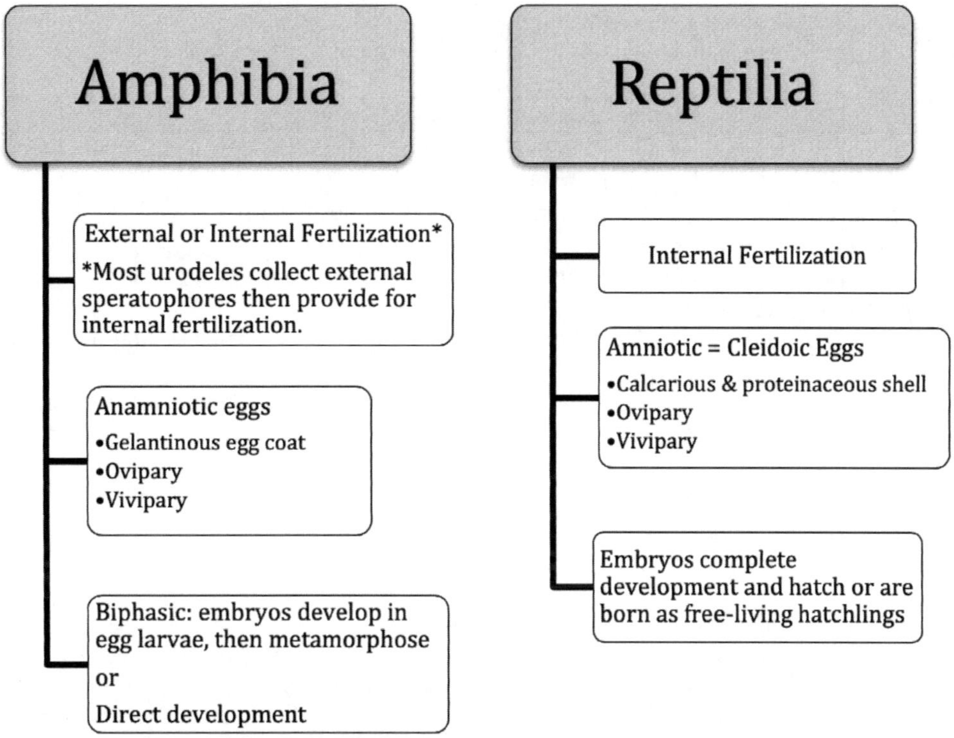


**Figure 1.** Fundamental characteristics of amphibian and reptilian reproduction.

## Amphibians

There is a rich literature on amphibian development including normal breeding and embryonic development. Much work focuses on a few frog and salamander species that serve as model species in developmental biological studies as well as those that are cultivated for the pet trade. These taxa include ranid frogs such as leopard frogs (*Rana pipiens)*, American bullfrogs (*Lithobates catesbeianus*, formerly *Rana catesbeianus*), European green frogs (*Pelophylax ridibus, P.* kl. *esculentus*), and old world African clawed frogs (*Xenopus laevis*). Salamander species include several *Ambystoma* species (large new world species like the tiger or spotted salamanders and the neotenic *Axolotl*), Eastern newts (*Notophthalmus viridescens*), European crested newts (*Triturus cristatus*). There is relatively little literature on caecilians and most focuses on particular aspects of three species (Table 1); however, recent studies of their reproductive behavior are increasing.[4]

**Table 1.** Amphibian taxa and links to literature and images of developmental stages. Additional references to specific stages by species are contained in many of the cited references in the text.

| Taxa | Information link |
|---|---|
| Axolotl *Ambystoma mexicanum* | https://embryology.med.unsw.edu.au/embryology/index.php/Axolotl_Development |
| Various anurans | https://www.jstor.org/stable/3890061 |
| African clawed frog *Xenopus laveus*<br>Leopard frog *Rana pipiens* | https://embryology.med.unsw.edu.au/embryology/index.php/Frog_Development |
| Leopard frog *Rana pipiens* | https://embryology.med.unsw.edu.au/embryology/index.php/Book_-_Normal_Plates_of_the_Development_of_Vertebrates_11 |
| Mudpuppy *Nectüriïs maculosus* | https://embryology.med.unsw.edu.au/embryology/index.php/Book_-_Normal_Plates_of_the_Development_of_Vertebrates_11 |
| African clawed frog *Xenopus laveus*<br>Dwarf clawed frog *Xenopus tropicus* | https://www.ibiology.org/development-and-stem-cells/frog-development/<br>(lecture) |
| Dendrobatid frogs (foam nesters) | https://www.pnas.org/content/104/29/11882 |
| Axolotl *Ambystoma mexicanum* | https://embryology.med.unsw.edu.au/embryology/index.php/Axolotl_Development |
| Four-toed salamander *Hemidactylium scutatum* | http://dx.doi.org/10.1016/j.mod.2014.12.007 |
| Crocodile newt *Echinotriton andersoni* | DOI: 10.3390/ani3030680 |
| Caecilians Ichthyophis kohtaoensis, I. *glutinosus*, *Hypogeophis rostratus*, and *Typhlonectes compressicauda* | https://doi.org/10.1002/(SICI)1097-4687(200001)243:1<3::AID-JMOR2>3.0.CO;2-D |

## Breeding patterns

Many amphibians, particularly in temperate regions, primarily breed in ephemeral wetlands. In tropical regions, reproduction often coincides with the rainy (monsoon) seasons. In both climates, because amphibians tend to rely on rainfall and species-appropriate temperatures to initiate breeding, suitable hydroregimes, (defined as water duration, timing, frequency, depth) are essential for reproductive success. In anurans (frogs and toads), breeding can be explosive and prolonged.[5] Explosive mating tends to occur in toads, calling is rare, and males greatly outnumber females. Males congregate and wrestle to access females. Generally fertilization rates are high. In prolonged mating (common in hylid and ranid frogs), males set up territories and call; females visit calling males and select a mate. Generally, males do not attempt to mate with other frogs passing through.[5] Gravid females extrude their eggs and males release sperm simultaneously onto the egg mass. Fertilization is external. Altig and McDiarmid (2007) provide thorough descriptions of egg types and egg mass structure.[6]

The diversity of Urodele courtship and mating postures are illustrated by Houck and Arnold (2003).[7] Urodeles (salamanders and newts) tend to have a form of internal fertilization in which the males deposit spermatophores (sperm packets) on the female's body or substrata, then the female collects and moves it into the cloaca; sperm are then housed in spermathecae (sperm storage organs) at the distal end of the oviducts.[7,8] Several species have external fertilization: Hynobiidae (Asiatic salamanders), Cryptobranchidae (giant salamanders including hellbenders), and Sirenidae (sirens). There are four common types of salamander egg incubation conditions: abandoning eggs in still waters, abandoning eggs in running waters, or direct development in terrestrial sites, often with egg guarding.[9,10]

Reproductive anatomy in amphibians includes secondary sex characteristics such as nuptial pads in anurans, sperm storage sacs of some females, mental glands of some salamanders, and copulatory appendages of some caecilians; however, most have a typical vertebrate system composed of paired gonads within the pleuroperitoneal cavity and gonadal ducts leading to the cloaca. In females the oviducal funnels receive ova, secretory glands within the oviduct coat each ovum with a jelly-layer. The egg jelly is necessary for sperm capacitation and polyspermy may occur because the development of sperm blockage is delayed about 30 minutes after fertilization. Wake and Dickie (1998) describe and discuss the diversity of amphibian oviducts.[11] Testes are elongated structures and are located in trunk. The cranial aspect of the testis contains spermatogonia surrounded by connective tissue and lobules with primary and secondary spermatocytes. The caudal aspect of the testis contains lobules with cysts of germinal cells in late stages of spermatogenesis, as spermatids and spermatozoa. Consequently urodele spermatogenesis is similar to that of fish and differs from amniote spermatogenesis.[12] The cranial kidney nephric tubules form a sexual segment such that nephric tubules also act as gonadal ducts as part of the epididymal complex in urodeles. Mature spermatozoa are transport to efferent ducts where they are stored until mating.[12]

## Development

Developing amphibian embryos tend to have radially symmetrical and holoblastic cleavage. Because yolk slows cleavage, the animal pole undergoes more rapid and complete cleavage. The embryos undergo a normal progression of stages following fertilization. Here I provide the most common developmental progressions.[13-15] Following fertilization, the embryo (now a zygote) undergoes cleavage with little or no growth to form a morula, blastula, gastrula, neurula, elongation and rotation, tailbud stage. Anurans, urodeles, and caecilians diverge such that most frogs and toads progress to early tadpole (hatchling), hind limb bud tadpole, forelimb metamorph, tail resorption stage, then froglet. In urodeles, the forelimb stage, hatching, hind limb stage, premetamorphic stage, metamorphed salamander. Some taxa have direct development such that the embryo develops through the larval stages and hatches with approximately adult form. There is no metamorphosis. In taxa with direct development, the sequence of ontogenetic stages between embryo and reproductive adult is abbreviated in comparison to that of related species.[16]

Most amphibians lay clutches of many eggs surrounded by gelatinous egg membranes. Eggs and early stage tend to be round with large yolks. Eggs may be pigmented, often at the animal pole. The gelatinous egg coat surrounding the oocyte, then the embryo does not prevent dehydration; consequently, eggs are often laid in water or on vegetation very near water. In terrestrial species, not associated with bodies of water, eggs are brooded by a parent.[17,18] Some anuran species have unique adaptation to allow brooding on or in their bodies[19] while other taxa attend their eggs using their bodies to prevent desiccation and perhaps prevent microbial infestation.[20,21] Hylid frogs in tropical regions may carry embryos on their backs to froglet stages; their embryos lack a tadpole stage. Other frogs brood their embryos in special sac-like structures in their backs (marsupial frogs) and others such as *Pipa* species, attach eggs to the dorsal skin and retain them until hatching. Several species of frogs create foam nests in which their embryos grow into larvae and hatch as tadpoles.[22,23]

Within the caecilians, a predominant number of species are viviparous. The oviparous species, *Ichthyophis kohtaoensis*, may lay eggs with developing embryos that range from various gastrula stages to neurula stages at deposition the clutch size can be small (4-6 eggs) in a direct-developing species, or large (~100 eggs) in other species.[24] Viviparious species tend to have small-medium sized clutches (20-50 eggs, summarized by Exbrayat (2006).[25]

The embryonic stages of *Ichthyophis kohtaoensis* as based upon a series of diagnostic features that include development of the lateral line organs, formation of three pairs of external gills, time and form of the developing eyes, changes in yolk structure, changes in the structure of the cloacal aperture, growth and changes in the tail, including formation and regression of the tail fin.[26] The developmental stages of viviparous caecilian species (*Typhlonectes compressicauda*) are described by Exbrayat (2006).[27] Embryonic nourishment source is taxon specific such that some are fully reliant on the yolk, others gain nutrition from uterine secretions, undeveloped oocytes, possibly dead embryos. Young from viviparous species are born just prior to metamorphosis or metamorphosed and may feed upon maternal skin.[28] Metamorphosis in caecilians is poorly described.

Parental care of eggs and offspring is relatively rare in amphibians. It is best described for Dendrobatid frogs in which both parents carry their offspring on their backs to water pools, and females feed young with trophic eggs. Parental care can be lengthy and costly to the parent; for example egg-brooding by plethodontid salamanders, and nutritional support to embryos (via specialized uterine secretions, oocytes, and skin to feed their larvae) in viviparous caecilian species. Crump (1996) summarized known parental care as "... widely distributed phylogenetically within the Amphibia, the percentage of species providing care is low.[29] Postoviposition parental care has been documented in only about 5% of caecilians, 18% of salamanders, and 6% of anurans. Internal gestation in the oviduct probably occurs in about 75% of all caecilians, in 4 species of salamanders, and in 5 species of anurans." These numbers have been slightly modified recently through further analysis of tropical anurans.[30]

## Reptilian Breeding Patterns

Reproductive cycles in reptiles can be quite variable. Consequently, not all species can be covered in this overview. Reproduction tends to be seasonal. Many reptile species found in captivity are from tropical regions. They tend to breed prior to or during the rainy season in the warmer spring and summer months. Some species from southern hemisphere may adjust breeding cycles to northern hemisphere while others retain some aspects of their native hemisphere reproductive cycles. In temperate species, mating is often in the spring with pregnancy then oviposition following in weeks or months. Some reptiles are known to store sperm, for months, or perhaps years, consequently reproduction may occur even when species have not mated recently. A few species of lizards are parthenogenic, and there are reports of facultative parthenogenesis in a several taxa of captive snakes.[31]

There is considerable diversity in the eggs and embryos of reptiles. Major clades differ in clutch sizes, egg structure, how embryos develop (oviparously, in a nest, or viviparously within the mother), the timing embryonic development, and timing of emergence. Eggs may have rigid shells, semirigid shells, or flexible (parchment) shells. The number and kinds of eggshell layers differs with taxa. The eggs of oviparous species have an outer mineral (calcareous) layer overlying organic fibrous layers (Fig 2).[32] The albumen is adjacent to the inside of fibrous layers (now termed the inner boundary of the shell membrane) when the egg starts development; as the embryo develops extra-embryonic membranes attach to the inner boundary. Eggs that undergo viviparous development have eggshell membranes that participate in various ways in the formation of a placenta.

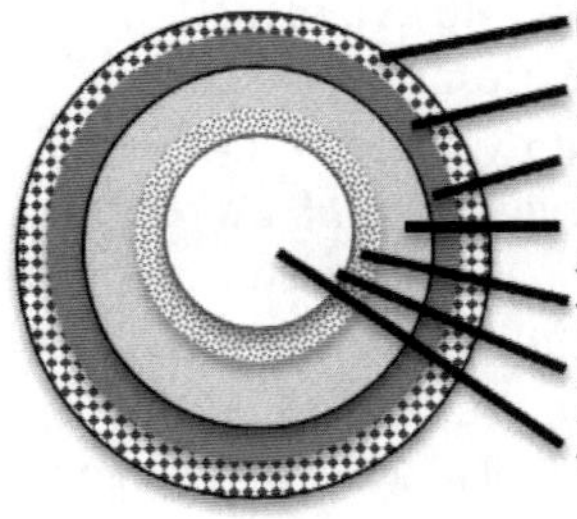


**Figure 2.** Cross-sectional diagram of reptile egg layers prior to chalking (Packard and DeMarco 1991).[32]

Within the Reptilia, the turtles, tortoises, lizards, and snakes all share similar early embryonic development. Fertilization is internal. All male reptiles, with the exception of the tuatara, have intromittent organs for sperm transfer (Sanger et al. 2015).[33] Tuataras rely upon cloacal apposition.[34]

Much of the recent development literature tends to focus on vivipary and temperature effects during incubation and sex determination. Several taxa are used as model organisms; consequently understanding of development is greater in several species of turtles (Testudinata), a few lizards and snakes (Lepidosauria), and a few species of Crocodilia. There is information on the tuatara (*Sphenodon punctatus*) recently assembled in one reference.[35] The reproductive biology of the fossorial Amphisbaenia (worm-lizards) is restricted to just a few papers that identify primarily oviparous forms as well as a few viviparous species.[36]

There is some diversity within breeding cycles, whether eggs undergo diapause or not, ranges of clutch sizes (1 to >140), whether eggs are laid or retained in the uterus so young are born, and in the form of embryonic nutrition. All are fundamental characteristics that are relevant to the health of both the mother and the successful development of the embryos. Most reptiles have terrestrial nests. Some gecko species attach their eggs individually to vegetation or structure. Brooding by many boid snakes species is common and eggs may adhere to one another in large clutches, while others snakes and lizards deposit clutches eggs that may barely contact one another. Eggs are generally white or cream colored but may become stained by soil or vegetation. Crocodilians and a few species of turtles build nest mounds and guard nests for at least the early part of development. Crocodilians provide parental care by assisting hatchings in escaping the nest.

## Development

Reptile eggs are provisioned with a large amount of yolk and relatively large offspring are produced. Similar to most amphibian zygotes, the dense yolk component of the zygote does not undergo mitotic divisions. Cells destined to contribute embryonic tissue are located at the edge of the yolk, a condition termed telolecithal.[37] A consequence is that early cell cleavage only can occur only in these cells. This developmental pattern, in which only a portion of the egg cell divides, is termed meroblastic cleavage and the early embryo, or blastodisk, forms on the surface of the yolk. A series of classical studies of early embryogenesis of reptiles established that gastrulation in turtles, lizards and snakes occurs via movement of cells through the blastopore, which is associated with a thickened part of the blastodisc. This blastopore pattern of gastrulation form is roughly similar to that of amphibians. It contrasts with that of birds and eutherian mammals that lack a blastopore. Their gastrulation occurs by ingression along an elongate thickening on the bastodisc, the primitive streak. Recent studies of development of turtles, Chinese softshell (*Pelodiscus sinensis*) and red-ear sliders (*Trachemys scripta*) and the helmeted chameleon lizard (*Chamaeleo calyptratus*) find that it is likely that gastrulation of the reptiles studies (turtles, lizards and snakes) likely are based upon shared processes with amphibians as well as with birds and eutherian mammals.[38,39]

The taxa differ greatly in how far development proceeds prior to oviposition. In turtles and tortoises, and some lizards, embryos undergo preovipositional developmental arrest (diapause) at the gastrula stage. Some remain in postovipositional arrest for weeks or months (reviewed by Rafferty and Reina, 2012). The recognition of normal developmental arrest can be assessed to identify viable eggs. Different taxa may undergo arrest or the embryos may die at different stages. In order to help distinguish between live and dead eggs non destructively, Rafferty and Reina (2012) identified several developmental events that can be assessed via egg candling (1) appearance of blood islands, (2) development of the haemodisc or vitelline circulation, (3) pigmentation of the eye, and (4) pigmentation of the body.[40] The appearance of chalking (white spot formation in turtle eggs) and banding (crocodilian eggs) have also been used as a defining some developmental stages. When turtle embryos resume development following early arrest, eggs visibly change color via the progressive formation of a chalky white area, termed white spot formation. The embryo and yolk are separated from the albumen by a vitelline membrane.[41] The albumen and yolk (with the embryo) segregate so that the yolk moves toward the egg's bottom (relative to gravity) and the vitelline membrane expands rapidly and displace albumen as it rises to occupy the upper portion of the egg. Where the vitelline membrane contacts and adheres to the eggshell membrane, chalking occurs.[41]

Embryonic development rates are temperature-dependent. The duration of incubation tends to be inversely related to incubation temperature. Embryos (not those in cold torpor, diapause, or developmental arrest) tend to develop more rapidly as temperatures increase, within the range of incubation temperatures that are normal for the species. At temperatures that are at warmer, but still support successful embryo growth, development tends to slow.[42] Above a certain threshold temperature (the thermal maximum) embryos die. Thermal maxima vary within taxa and within species across latitudes.

Most turtle species, some lizards, crocodilians, and the tuatara have environmentally determined sex (temperature dependent sex determination or TSD).[43] In species with TSD, the incubation environment directs the embryo's sex. Turtles have one of two types of TSD: warm female-cool male or hot and cool male with intermediate temperatures producing females. Crocodilians and some lizards have this latter pattern. The tuatara has a cool female-warm male response system. Those species with chromosomal sex determination have independently evolved heteromorphic (XY) sex chromosomes in several distinct lineages (South American chelids [*Acanthochelys*], Mexican musk turtles [*Saurotypus*], and Geoemydid species from Southeast Asia and the Philippines in the genus *Siebenockiella*). Several other lineages have homomorphic sex chromosomes (Australian pleurodires *Emydura* and *Chelodina*, softshelled turtles in the genera *Pelodiscus* and *Apalone*, the narrow-bridged musk turtle *Claudius* from Central America, the wood and bog turtles of North America now *Glypemys*).[43] Bearded dragons *Pogona vitticeps* have both a chromosomal system and TSD system.[44] All snakes have chromosomally determined sex.[43]

Standard or normal stages series are fundamental tools for assessing development. They are based upon and delineate particular developmental events that occur sequentially. Most standard staging series are collected at constant temperatures so that the sequence of morphological changes can be used comparatively to identify developmental status, at any temperature tolerated by the embryos. Developmental stages beyond gastrulation have been established for several species of testudines, lepidosaurs, and crocodilians. Developmental stages are helpful in assessing developmental progress as well as failures. Table 2 summarizes both species and citations for currently available series. Generally, the species represented in these series do not have prolonged periods of developmental arrest. Consequently, major taxa of turtles, tortoises, lizards and snakes are not represented. Most series start with post-ovipositional stages.

**Table 2.** Summary of normal developmental staging series for select reptiles.

| Reptile | Literature source |
|---|---|
| European pond turtle (*Emys orbicularis*) | Mahmoud et al. 1973[45] |
| Common snapping turtle (*Chelydra serpentina*) | Yntema 1968[46] |
| | reproduced in Ewert 1985[41] |
| Red-eared slider (*Trachemys scripta*) | Greenbaum 2002[47] |
| Painted turtle (*Chysemys picta*) | Mahmoud et al. 1973[45] |
| | Cordero & Janzen 2014[48] |
| Pink-bellied side neck turtle *Emydura subglobosa* | Werneburg et al. 2009[49] |
| Chinese softshell (*Pelodiscus sinensis*) | Tokita and Kuratani 2001[50] |
| Spiny softshell *Apalone spinifera* | Greenbaum and Carr 2002[51] |
| Fly River turtle *Carettochelys insculpta* | Beggs 2000[52] |
| Sea turtles (all species) | Miller 1985[53] |
| Olive ridley turtle *Lepidochelys olivacea* | Crastz 1982[54] |
| Crocodilians *Alligator*, Crocodylis *niloticus, C. johnsoni,* | Ferguson 1985[55] |
| *Caiman latirostris* | Iungman et al. 2008[56] |
| Tuatara *Sphenodon puctatus* | Dendy 1899[57] |
| Green iguana *Iguana iguana* | Lima et al. 2018[58] |
| Anoles *Anolis sagrei* | Sanger et al. 2008[59] |
| Leopard gecko *Eublepharis macularis* | Wise et al. 2009[60] |
| Madagascar ground gecko *Paroedura pictus* | Noro et al. 2009[61] |
| South American collared lizard *Tropidurus torquatus* | Rapp Py-Daniel et al. 2017[62] |
| Oriental garden lizard *Calotes versicolor* | Muthukkaruppan et al. 1970[63] |
| Japanese striped snake, *Elaphe quadrivirgata* | Matsubara et al. 2014[64] |
| African house snake *Boaedon fuliginosus* | Boback et al. 2012[65] |
| Egyptian cobra *Naja h. haje* | Khannoon and Evans 2014[66] |
| African rock python *Python sebae* | Boughner et al. 2007[67] |
| Garter snake *Thamnophis sirtalis sirtalis* | Zehr 1962[68] |

Because of the diversity of amphibian and reptilian reproductive patterns and incubation requirements, specific information about taxa should be sought from those that work with that species in the field or in captivity.

## References


1. Wake DB. Ecology. Facing extinction in real time. *Science.* 2012;335(6072):1052-1053. doi:10.1126/science.1218364.

2. Blackburn DG, Starck JM. Morphological specializations for fetal maintenance in viviparous vertebrates: an introduction and historical retrospective. *J Morphol.* 2015;276(11): E1–E16. doi:10.1002/jmor.20410.

3. Blackburn DG. Evolution of vertebrate viviparity and specializations for fetal nutrition: a quantitative and qualitative analysis. *J Morphol.* 2015;276(8):961-990.

4. Kupfer A, Maxwell E, Reinhard S, Kuehnel S. The evolution of parental investment in caecilian amphibians: a comparative approach. *Biol J Linn Soc.* 2016;119(1):4-14.

5. Wells KD. The social behavior of anuran amphibians. *Anim Behav.* 1977;25:666-693.

6. McDiarmid RW, Altig R. Research methods. In: McDiarmid RW, Altig R, eds. *Tadpoles. The Biology of Anuran Larvae.* Chicago, IL: University of Chicago Press;2000.

7. Houck LD, Arnold SJ. Courtship and mating behavior. In: Jamieson BGM, ed. *Reproductive Biology and Phylogeny of Urodela.* Boca Raton, FL: CRC Press;2003:383-342.

8. Salthe SN. Courtship patterns and the phylogeny of the urodeles. *Copeia.* 1967:100-117.

9. Salthe SN. Reproductive modes and the number and sizes of ova in the urodeles. *Am Midland Nat.* 1969; 467-490.

10. Salthe SN, Mecham JS. Reproductive and courtship patterns. *Physiol Amphibia.* 1974;2:309-521.

11. Wake MH, Dickie R. Oviduct structure and function and reproductive modes in amphibians. *J Exp Zool.* 1998;*282*(4-5):477-506.

12. Uribe Aranzábal MC. The testes, spermatogenesis male reproductive ducts. In: Jamieson BGM, ed. *Reproductive Biology and Phylogeny of Urodela.* Boca Raton, FL: CRC Press;2003:183-2002.

13. Bernabò I, Brunelli E. Comparative morphological analysis during larval development of three syntopic newt species (Urodela: Salamandridae). *Eur Zool J.* 2019;86(1):38-53.

14. Hurney CA, Babcock SK, Shook DR, et al. Normal table of embryonic development in the four-toed salamander, *Hemidactylium scutatum. Mech Dev.* 2015;136: 99-110.

15. Harrison RG. Harrison stages and description of the normal development of the spotted salamander, *Amblystoma punctatum* (Linn). In: Wilens S, ed. *Organization and Development of the Embryo.* New Haven, CT: Yale University Press;1969:44-66.

16. Hanken JA. Direct development. In: Hall BK, Olson WM, eds. *Keywords and Concepts in Evolutionary Developmental Biology*. Cambridge, MA: Harvard University Press;2003:97-102.

17. Woodruff DS. Male postmating brooding behavior in three Australian Pseudophryne (Anura: Leptodactylidae). *Herpetologica.* 1977;33(3):296-303.

18. Taigen TL, Pough FH, Stewart MM. Water balance of terrestrial anuran (*Eleutherodactylus coqui*) eggs: importance of parental care. *Ecology.* 1984;65(1):248-255.

19. Elinson RP, del Pino EM, Townsend DS, et al. A practical guide to the developmental biology of terrestrial-breeding frogs. *Biol Bull.* 1990;179(2):163-177.

20. Forester DC. Brooding behavior by the mountain dusky salamander: can the female's presence reduce clutch desiccation? *Herpetologica.* 1984;105-109.

21. Highton R, Savage T. Functions of the brooding behavior in the female red-backed salamander, *Plethodon cinereus*. *Copeia.* 1961;1:95-98.

22. Elinson R, del Pino EM. Developmental diversity of amphibians. *Dev Biol.* 2012;1(3):345-369.

23. Crump ML. Amphibian diversity and life history. In: Dodd Jr CK, ed. *Amphibian Ecology and Conservation. A Handbook of Techniques.* New York, NY: Oxford University Press;2009:3-20.

24. Kupfer A, Kramer A, Himstedt W, Greven, H. Copulation and egg retention in an oviparous caecilian (Amphibia: Gymnophiona). *Zoologischer Anzeiger-A J Comp Zool.* 2006;244(3-4):223-228.

25. Exbrayat J-M. Oogenesis and Folliculogenesis. In: Jamieson BGM, ed. *Reproductive Biology and Phylogeny of Gymnophiona: Caecilians*, Boca Raton, FL: CRC Press;2006a:276-290.

26. Dünker N, Wake MH, Olson WM. Embryonic and larval development in the caecilian *Ichthyophis kohtaoensis* (Amphibia, Gymnophiona): a staging table. *J Morphol.* 2000;243(1):3-34.

27. Exbrayat J-M. Fertilization and embryonic development. In: Jamieson BGM, ed. *Reproductive Biology and Phylogeny of Gymnophiona: Caecilians.* Boca Raton, FL: CRC Press;2006b:359-386.

28. Wake MH. A brief history of research on gymnophionan reproductive biology and development. In: Jamieson BGM, ed. *Reproductive Biology and Phylogeny of Gymnophiona: Caecilians.* Boca Raton, FL: CRC Press;2006:1-38.

29. Crump ML. Parental care among the amphibia. In: *Advances in the Study of Behavior.* New York, NY: Academic Press;1996:(25)109-144.

30. Summers K, Tumulty J. Parental care, sexual selection, and mating systems in neotropical poison frogs. In: Macedo RH, Macedo G, eds. *Sexual Selection.* New York, NY: Elsevier;2014:289-320.

31. Booth W, Schuett GW. The emerging phylogenetic pattern of parthenogenesis in snakes. *Biol J Linn Soc.* 2016;118(2):172-186.

32. Packard MJ, DeMarco VG. Eggshell structure and formation in eggs of oviparous reptiles. In: Deeming DC, Ferguson MWJ, eds. *Egg Incubation: Its Effects on Embryonic Development in Birds and Reptiles.* New York, NY: Cambridge University Press;1991:53-69.

33. Sanger TJ, Gredler ML, Cohn MJ. Resurrecting embryos of the tuatara, *Sphenodon punctatus*, to resolve vertebrate phallus evolution. *Biol Lett.* 2015;11:20150694. http://dx.doi.org/10.1098/rsbl.2015.0694

34. Gans C, Gillingham JC, Clark DL. Courtship, mating and male combat in tuatara, *Sphenodon punctatus*. *J Herpetol.* 1984;18(2):194-197.

35. Cree A. *Tuatara: Biology and Conservation of a Venerable Survivor*. Christchurch, NZ: Canterbury University Press;2014.

36. Andrade DV, Nascimento LB, Abe AS. Habits hidden underground: a review on the reproduction of the Amphisbaenia with notes on four neotropical species. *Amphibia-Reptilia.* 2006;27(2):207-217.

37. Balfour FM. *A Treatise on Comparative Embryology.* Vol. 1. London, UK: Macmillan Publishers Limited;1880.

38. Stower MJ, Diaz RE, Fernandez LC, et al. Bi-modal strategy of gastrulation in reptiles. *Dev Dyn.* 2015;244(9):1144-1157.

39. Stower MJ, Bertocchini F. The evolution of amniote gastrulation: the blastopore-primitive streak transition. *Dev Biol.* 2017;6(2):e262.

40. Rafferty AR, Reina RD. Arrested embryonic development: a review of strategies to delay hatching in egg-laying reptiles. *Proc Royal Soc London B: Biol Sci.* 2012;doi:10.1098/rspb.2012.0100.

41. Ewert MA. Embryology of turtles. In: Gans C, Billet F, Maderson P, eds. *Biology of the Reptilia.* New York, NY: John Wiley & Sons;1985:75-267.

42. Bowden RM, Carter AW, Paitz RT. Constancy in an inconstant world: moving beyond constant temperatures in the study of reptilian incubation. *Integr Comp Biol.* 2014;54(5):830-840.

43. Janzen FJ, Phillips PC. Exploring the evolution of environmental sex determination, especially in reptiles. *J Evol Biol.* 2006;19(6):1775-1784.

44. Deveson IW, Holleley CE, Blackburn J, et al. 2017. Differential intron retention in Jumonji chromatin modifier genes is implicated in reptile temperature-dependent sex determination. *Sci Adv.* 2017;3(6):e1700731.

45. Mahmoud IY, Klicka J, Hess GL. Normal embryonic stages of the western painted turtle, *Chrysemys picta bellii. J Morphol.* 1973;https://doi.org/10.1002/jmor.1051410303.

46. Yntema C. A series of stages in the embryonic development of *Chelydra serpentina. J Morphol.* 1968;125(2):219-251.

47. Greenbaum E. A standardized series of embryonic stages for the emydid turtle *Trachemys scripta. Can J Zool.* 2002;80(8):1350-1370.

48. Cordero G, Janzen FJ. An enhanced developmental staging table for the painted turtle, *Chrysemys picta* (Testudines: Emydidae). *J Morphol.* 2014;275(4):442-55.

49. Werneburg I, Hugi J, Müller J, Sánchez-Villagra MR. Embryogenesis and ossification of *Emydura subglobosa* (Testudines, Pleurodira, Chelidae) and patterns of turtle development. *Dev Dyn.* 2009;238(11):2770-2786.

50. Tokita M, Kuratani S. 2001. Normal embryonic stages of the Chinese softshelled turtle *Pelodiscus sinensis* (Trionychidae). *Zool Sci.* 2001;18:705-715.

51. Greenbaum E, Carr JL. Staging criteria for embryos of the spiny softshell turtle, *Apalone spinifera* (Testudines: Trionychidae) *J Morphol.* 2002;254:272-291.

52. Beggs K, Young J, Georges A, West P. Aging the eggs and nest of the pig-nosed turtle (*Carettochelys insculpta*) from Northern Australia. *Can J Zool.* 2000;78:373-392.

53. Miller JD. Embryology of marine turtles. In: Gans C, Billett F, Maderson PFA, eds. *Biology of the Reptilia.* Vol 14A. New York, NY: Wiley-Interscience;1974:269-328.

54. Crastz F. Embryological stages of the marine turtle *Lepidochelys olivacea* (Eschscholtz). *Revista de Biologia Tropical*. 1982;30(2):113-120.

55. Ferguson MWJ. Reproductive biology and embryology of the crocodilians. In: Gans C, Billet F, Maderson PFA, eds, *Biology of the Reptilia.* New York, NY: John Wiley & Sons;1985:329-491.

56. Iungman J, Piña CI, Siroski P. Embryological development of *Caiman latirostris* (Crocodylia: Alligatoridae). *Genesis.* 2008;46:401-417.

57. Dendy A. Memoirs: Outlines of the Development of the Tuatara, *Sphenodon* (*Hatteria*) *punctatus*. *J Cell Sci.* 1899;2(165):1-87.

58. Lima FC, Py-Daniel TR, Sartori MR, et al. Developmental staging table of the green iguana. *Acta Zoologica*. 2018:1-13.

59. Sanger TJ, Losos JB, Gibson-Brown JJ. A developmental staging series for the lizard genus *Anolis*: a new system for the integration of evolution, development, and ecology. *J Morphol.* 2008;269:129-137. doi:10.1002/jmor.

60. Wise PAD, Vickaryous MK, Russell AP. An embryonic staging table for in ovo development of *Eublepharis macularius*, the leopard gecko. *Anat Rec.* 2009;292:1198-1212. doi:10.1002/ar.20945.

61. Noro M, Uejima A, Abe G, et al. Normal developmental stages of the Madagascar ground gecko *Paroedura pictus* with special reference to limb morphogenesis. *Dev Dyn.* 2009;238:100-109.

62. Rapp Py-Daniel T, Kennedy Soares De-Lima A, et al. A staging table of post-ovipositional development for the South American collared lizard *Tropidurus torquatus* (Squamata: Tropiduridae). *Anat Rec.* 2017;300(2):277-290.

63. Muthukkaruppan V, Kanakambika P, Manickavel V, Veeraraghavan K. Analysis of the development of the lizard, *Calotes versicolor* I. a series of normal stages in the embryonic development. *J Morphol.* 1970;130:479-490. doi:10.1017/S1355617712001555.

64. Matsubara Y, Sakai A, Kuroiwa A, Suzuki T. Efficient embryonic culture method for the Japanese striped snake, *Elaphe quadrivirgata*, and its early developmental stages. *Dev Growth Differ*. 2014;56(8):573-82.

65. Boback SM, Dichter EK, Mistry HL. A developmental staging series for the African house snake, *Boaedon (Lamprophis) fuliginosus*. *Zoology*. 2012;115(1):38-46.

66. Khannoon ER, Evans SE. The embryonic development of the Egyptian cobra *Naja h. haje* (Squamata: Serpentes: Elapidae). *Acta Zoologica*. 2014;95(4):472-83.

67. Boughner JC, Buchtová M, Fu K, et al. Embryonic development of *Python sebae*–I: Staging criteria and macroscopic skeletal morphogenesis of the head and limbs. *Zoology*. 2007;110(3):212-230.

68. Zehr DR. Stages in the normal development of the common garter snake, *Thamnophis sirtalis sirtalis*. *Copeia.* 1962:322-329. doi:10.2307/1440898.

# Diagnosis via the Herptile Egg

J. Jill Heatley, DVM, MS, DABVP (Avian Practice), DABVP (Reptile and Amphibian Practice), Dipl ACZM

**Session #2791**

*Affiliation:* From the Department of Small Animal Clinical Sciences, College of Veterinary Medicine and Biomedical Sciences, Texas A&M University, College Station, TX 77843, USA.

## Introduction

This manuscript aims to provide a framework for evaluation of the reptile wherein reproduction has not occurred. Reptile and amphibian eggs are seldom presented to the veterinarian for diagnosis until a problem—usually a medical or surgical problem in a female—is identified. Limited presentation is likely due to many factors including lack of basic diagnostic guidelines, lack of easily accessible normal values for many species, and lack of veterinary involvement in reptile and amphibian reproduction in the pet sector. Comments here are restricted to reptiles that are oviparous or ovoviviparous, or the egg producing species. The reptile egg is much more resistant to desiccation than amphibian eggs and the relatively large yolk provides energy and maternal immunoglobulins that may last up to a year post hatch. The development of the amniotic eggs is key to the classification of reptiles as such.

## Sample Handling and Shipment

You will get the most out of an egg necropsy by doing it yourself and thorough image documentation. As in other necropsies, the fresher the carcass, the greater the diagnostic yield. Refrigerating the egg prior to necropsy will reduce bacterial and fungal growth and therefore reasonably preserve the carcass. Freezing will also reduce odor as well as bacterial and fungal growth and load, but may limit histopathologic diagnosis or irritate the pathologist. Refrigerated eggs are better preserved for cytology, while culture and PCR diagnostics may be better served by freezing.

## Storage and Shipping of Dead Eggs

Place in 3 large plastic zipper lock or leak-proof bags, each sealed with packing tape. Inside the second leak-proof plastic bag, place sufficient absorbent material and freezer packs. Place paperwork in a separate, sealed plastic bag and label all samples. Use a permanent marker on the internal-most bag and pencil on waterproof paper in the second-layer bag. I prefer to label eggs with a Sharpie marker; reasonable alternatives are graphic pencil or china marking (wax) pencil. In the refrigerator, also provide eggs triple enclosure with the outer layer a hard plastic, odor proof enclosure (glad container or equivalent) to avoid crush. For shipping, place the paperwork and carcass in a shipping container with an insulated outer container, which should be crush-proof. Large plastic picnic coolers are ideal and do not need to be placed inside a cardboard box for shipping. Cardboard boxes without rigid coolers are inappropriate because condensation causes them to degrade in shipping. Ship overnight on cold packs.

## Preliminary Pre-Incubation Assessment of Eggs

A wide range of normality makes pre-incubation assessment challenging, and few studies have provided normal expected specifications of the healthy reptile and or amphibian eggs. Laid eggs may be collected as a clutch or separately depending on the species and may be collected directly from the female or collected later from the nest. We do not know what effect, if any, human normal dermal flora may have on reptile eggs, and salmonellosis and leptospirosis have been associated with reptiles eggs.[1] Gloves are recommended for handling reptile eggs. Consider documenting the following reptile egg characteristics: shape, surface texture and color, weight, shell pliability.

### Egg length, width, and shape

These characteristics may be highly variable in some reptile species without apparent effects on hatchability or survivability. An accurate small gram scale with a bowl-like weight platform and calipers should be used. More research and documentation of the diseased egg is needed or hatchling death associated with eggs shape and size are needed. Unlike avian species, differing sizes and shapes are not yet specifically associated with increased morbidity or mortality in reptiles. With the exception of certain geckos and some chelonians, eggs of oviparous reptiles absorb water from the substrate during the course of normal incubation. As the rate of water absorption exceeds the rate of water loss by transpiration from exposed surfaces, the eggs swell during incubation.[2]

### Egg surface texture and color

To my knowledge all healthy reptile eggs are white to off-white to tan colored. This color may change during incubation, likely associated with the substrate. Birds are the only living amniotes with colored eggs; however, a recent study demonstrated red-brown protoporphyrin IX and blue-green biliverdin—the pigments responsible for variation in avian egg color—in fossilized eggshell of a nonavian dinosaur. So modern birds may have inherited egg color from their nonavian dinosaur ancestors, or egg color could have evolved independently multiple times.[3,4] In monitors and other species with pliable eggs (squamates), egg shape and surface texture may indicate dehydration, euhydration or hyperhydration of the egg. A dry reptile egg may collapse or dimple, often early in development, and is an indication to increase incubation moisture. Excess moisture may be absorbed into the egg often toward the end of incubation. Although challenging to recognize, the egg may appear bloated or distorted in the worst cases. Should excess moisture be drawn inside, the embryo may drown. In species with nonpliable eggs, one would expect that, as in birds, a slow weight change might occur during incubation. To my knowledge this has not been documented in a reptilian species and may be confounded in pliable egg species based on absorption of moisture from the incubation substrate or environment. While one author has stated "Discolored areas on the shell may represent colonies of harmless bacteria and fungi is generally not worth trying to control unless they become overwhelming," I would consider candling and moving or separating eggs that are suffering fungal or bacterial growth because they are a danger to other eggs. Another author states "Secretion of substance upon the shell surface is immediate indication of microbial invasion and cause for [egg] removal from the incubator."[5] The urgency and negative connotation of some of these appearances likely is species based: both fungal and microbial as well as reptilian in nature.

### Egg weight

Egg weight has significant effects on reptile hatchability and neonate survivability (Table 1) although time of year of hatch may play a more significant role in survival of squamates. Bigger heavier eggs fare better than smaller eggs. Minimal data are available for egg weights because most breeders do not handle the egg to avoid damaging embryo adherence.

### Egg shell pliability

Most tortoise eggs and some crocodile and lizard eggs are rigid, while turtle, snake, lizard, crocodilian eggs tend to be a pliable, leathery, or parchment-like texture. Many reptile species have a wide range of egg shape, size, and weight within normal limits, even within a single clutch. The delicate nature of amphibian eggs suggests they are not separated or handled; however, an observational exam should occur, possibly using a dissecting microscope and a wet-weight of the egg mass along with an enumeration of the eggs.

As in avian species, reptile eggs are generally handled minimally during the first trimester to avoid addling or disruption of early embryonic growth (embryonic vessel and disc) and death. In general, Chelonia and Crocodylia (as archosaurs) have rigid, heavily mineralized shells. Some turtles are characterized by pliable "parchment" shelled eggs with only a thin mineral layer (Table 1). Shell mineral content has a physiologic correlation: calcium source for the embryo in the archosaur egg is mainly the shell, while in squamates it is mainly the yolk. *Iguana iguana* is intermediate with ~50% calcium provided by the shell. Water exchange also is affected by shell type and is negligible in hard-shelled eggs while pliable eggs generally vary in water uptake, which in snakes is necessary for successful development. In *Chelydra serpentina*, loss or gain of up to 30% mass may occur.

**Table 1.** Exemplar egg parameters for select reptile species.

| Species | Egg type | Egg mass (g) | Incubation duration (days) |
|---|---|---|---|
| *Chelydra serpentine* | Parchment | 12.9 | 65.6 |
| *Uromastyx aegyptius* | Parchment | 17.9 | 74 |
| *Varanus komodensis* | Parchment | 175 | 236 |
| *Testudo graeca* | Rigid | 16 | 76 |
| *Trionyx triunguis* | Rigid | 78.2 | 82 |
| *Hemidactylus turcicus* | Rigid | 0.4 | 46 |

### Candling

Candling is most often used to assess 1) egg viability in the first eggmester via assessment of formation of vessels and embryonic disk and 2) continuing viability in the second and third trimesters based on continuing draw down, air cell viability and later embryo movement. A general recommendation for handling reptile eggs is no more often than once weekly to monthly (once to twice per eggmester) to avoid early embryonic death. Use a cool light-source for candling (eg, Remington Bore Light, flexible LED, Interstate Arms Corp, Billerica, MA 01821, USA) to avoid harm to developing embryos.

You can distinguish viable reptile eggs from infertile reptile eggs by candling them in species that don't go into a resting period (diapause) after laying. In general, species that lay thinner-shelled pliable eggs (most squamates) typically do not use developmental arrest after oviposition and as a result they have shorter incubation periods than those laying thicker, more rigid, brittle-shelled eggs that do use it (Table 2).[6] Monitor lizard eggs generally develop quickly, absorbing moisture from the substrate, exchanging gases and growing. The growing reptile embryo develops blood vessels and tissues on inner surface of the shell, the embryonic disk. Example of expectations for candling eggs the monitor lizards[7]: The infertile reptile egg remains clear based on lack of growth. A week after laying, the embryo of a fertile egg begins to grow as evidenced by a small round disk. The following week later, disk has grown, and more of the shell becomes opaque, as the light fails to penetrate the tissues and blood vessels. After several weeks, the entire egg appears opaque, as the body of the embryo and its many tissues stops all light. In avian embryos this is when I look for the movement of an extremity.

As an example in red-footed tortoises, I collect, weigh, and set eggs in the incubator and create records on day 1 of discovery. This year I waited until April, but I suspect eggs were laid as early as January or February. Egg Buddy candling is scheduled for 6 weeks after the start of incubation. This timing allows for discrimination of live and dead eggs and for the main goal of removal dead eggs, which are excellent growth media for infectious diseases (eg, virus, bacteria, fungus, and maybe *Coccidia*), with a naturally prolonged incubation expected of from 141-185 days (up to 200, 50-day span recorded within clutches). Great variation occurs in incubation time among reptile species (Table 1).

You may move reptile eggs a few hours after deposition prior to white patch or chalking: prior to the yolk sac membrane growing and attaching itself to the surface of the egg shell. Chilling eggs to 5-15 °C immediately after oviposition can prolong this initial period of safe movement. Unlike in avian species, the reptile embryo adheres to the inside of the egg creating a "white patch." This site should be noted during egg collection and remain up. Thus most reptile eggs are incubated in the orientation at which they were laid and are not turned. Excessive movement of the egg after white patch appearance can shear the connection of the embryo and internal shell resulting in fatality. Rigid-shelled egg species (turtles) and crocodilians) appear more sensitive to addling of the egg. Upon white patch expansion to 50%, gentle egg movement is acceptable, and upon encompassment of the entire egg embryos are generally robust and tolerant to jarring and rotation; however, some squamate and turtle species may tolerate or even benefit from turning (*E. macularius*, *Elaphe guttata*, *Python molurus*, *Chrysemys picta*, alligators) but with a few lizard exceptions (*Eumeces* spp., *Anolis carolinesis*) turning of eggs does not occur in the wild.

**Table 2.** Embryonic stage at which reptile eggs are laid varies among reptile types.

| Embryonic stage | Reptile | During egg retention |
|---|---|---|
| Gastrulation | Turtles, *Sphenodon* chameleons (*C. calyptratus, Furcifer lateralis*) | Arrested at gastrula stage |
| Neurulae | Crocodilians | Egg develops internally, abnormally if retained |
| Varies, often limb bud stage | Most squamates | Most squamates retain eggs for about 1/4 of embryogenesis |
| Late embryonic development | Yellow-bellied 3-toed skink (*Saiphos equalis*) (AU) Smooth green snake (*Opheodrys vernalis*) (US) | |

## Other egg assessments

The Egg Buddy (Cornwall, UK, www.avitronics.co.uk) also may be used for reptile eggs. If using Buddy for reptile eggs you can expect heart rates ~ 60-70 BPM. This is a digital laser assessment of pulse and heart rate. For early assessment (first trimester) eggs should be placed in the Buddy maintaining the orientation that they were laid. Endoscopy other egg interventions have received minimal study likely based on the prolonged time that some species may loll about the shell without apparent problems. Pipped red-footed tortoises may remain in the shell for several days.[8] Overzealous hatch intervention is to be avoided as disruption of egg membrane, umbilical or yolk vessels or yolk internalization process may result in serious injury to the hatchling.

## Egg necropsy

Egg necropsy is best performed in a similar manner for each and every egg for best results. Use of a standardized protocol and record sheet is advised (Table 3). In general, egg measurements are following by opening, microbial sampling, and embryonic staging. The embryo may be further sampled but is often best placed in formalin for histopathologic exam. The yolk and membranes are seldom saved but maybe frozen for microbial or toxicologic assay. A perivitelline membrane assay may be performed to assess sperm penetration of the ovum (egg fertilization).

**Table 3.** Reptile egg necropsy record.

Egg Identification: Species:

| | | | |
|---|---|---|---|
| Date of egg | Collection | Incubation start | Necropsy |
| | | | |
| Egg measures | Weight (g) | Length (cm) | Width (cm) |
| | | | |
| Egg shape | Symmetrical | Spheroid | Oblate/ |
| | Asymmetrical | Ellipitical/oblong | Photo |

| | | | | |
|---|---|---|---|---|
| Egg imaging method | Candle | Radiograph/fluoro | CT | *Photo* |
| Note if present | Air cell | Clear mobile yolk | Free mobile gas | *Photo* |
| | Crack(s) | Shell imperfection | Blood crescent or ring | |
| Intact egg diagram<br>Notes | | | | |

| | | |
|---|---|---|
| Break out: Avoid egg contamination, sample(s). 1) Culture, 2) PCR, 3) Cytology/virology. | | |
| Samples obtained<br>Describe | | Bacteriology |
| | | Mycology |
| | | PCR |
| | | Cytology |
| | | Virology |
| Infertile | Fertile | Photo |
| Early Embryo | Disk or vessels present | |
| Mid Embryo | Identifiable notochord, heart eye spot, no limb buds | |
| Late Embryo | Describe embryo position<br>Egg tooth present?<br>Lungs inflated? | |
| Break out<br>Notes<br>Diagram | | |

## Embryonic Staging

In avian medicine, embryonic stage determination for the dead fetus provides a framework for differentials of mortality etiology (Table 4). Embryonic staging has been determined for many squamates, but for few pet species. I recommend embryonic staging based on four broad categories: early embryonic (eggmester I), midembryonic (eggmester II), late embryonic (eggmester III, synonymous with failure to internal or external pip and failure to hatch), and failure to finish hatch/exit/difficult emergence (eggmester IV). Although embryonic staging has been performed for many reptiles, it is rarely applicable in exotic animal practice. Variable incubation times and development rates of reptile embryos may have evolved to allow synchrony of the hatchling group to dilute predation risk upon hatchling emergence.[9] In the Australian freshwater turtle, *Emydura macquarii*, less advanced embryos within a clutch either accelerated their development or they hatched prematurely (by up to 5 days) to allow synchrony of hatching and hatchling group formation for emergence.

**Table 4.** Select examples of embryonic staging available for or applicable to exotic pet reptiles.

| Genus species | Common name | Embryonic staging | Reference |
|---|---|---|---|
| *Eublepharis macularius* | Leopard gecko | Post oviposition | Wise 2009[a] |
| *Python sebae* | Rock python | Post oviposition | Boughner 2007[b] |
| *Chamaeleo lateralis* | Jewel chameleon | Pre and post oviposition | Blanc 1974[c] |
| *Chamaeleo calyptratus* | Veiled chameleon | | Diaz 2017[d] |
| *Hemidactylus turcicus* | Gecko | Post oviposition | Werner 1971[e] |

[a]Wise PA, Vickaryous MK, Russell AP. An embryonic staging table for in ovo development of *Eublepharis macularius*, the leopard gecko. *Anat Rec.* 2009; 292(8):1198-1212.
[b]Boughner JC, Buchtová M, Fu K, et al. Embryonic development of *Python sebae* - I: Staging criteria and macroscopic skeletal morphogenesis of the head and limbs. *Zoology (Jena).* 2007;110(3):212-230.
[c]Blanc F. 1974. Table de de´veloppement de *Chamaeleo lateralis* Gray, 1831. *Ann d'Embryol Morphogen.* 5:99-115.
[d]Diaz RE, Bertocchini F, Trainor PA. Lifting the veil on reptile embryology: the veiled chameleon (*Chamaeleo calyptratus*) as a model system to study reptilian development. In: *Avian and Reptilian Developmental Biology.* New York, NY: Humana Press;2017:269-284.
[e]Werner YL. The ontogenic development of the vertebrae in some Gekkonoid lizards. *J Morphol.* 1971;133(1):41-91.

### Hatching and emergence

The egg tooth of chelonia and crocodylia is not a true tooth but a horny outgrowth at the midline of the upper jaw; however, the egg tooth of squamates is a true dentinal tooth on the premaxilla. Reptile behavior at hatching appears poorly determined and documented in the literature but anecdotal evidence abounds on the internet. Eggs of Crocodilia and Chelonia, have a pair of egg membranes separating a thick layer of albumen from the calcareous shell. In contrast, eggs of oviparous Lepidosauria have only a single shell membrane. The volume of albumen in squamate eggs is relatively small at the time of oviposition.

For the parchment-like shell, hatchlings slit or tear multiple wide arcs using the egg tooth or caruncle. The claws on the forelimbs of turtles may help tear open the shell. Fluid within the egg is necessary for hatching and is released during the hatching process. Should excess fluid drain from the egg, the shell membranes dry, hatchlings may become trapped in the shell and die. The hatching process ranges from hours to several days. An expected internal pipping time and egg exit interval have not been specifically documented in most reptiles. Hatchling tortoises must remain within the shell as necessary to allow yolk sac internalization and resorption of egg membrane vessels.

## Infectious Diseases and Sampling of the Egg

Infection prevention in the incubator is important to avoid microbial infection and death of embryos.[5] In the loggerhead sea turtle, *Caretta caretta* a high species diversity of bacteria within eggs or the occurrence of the same bacteria in both females and their eggs was correlated with lower hatching success. Salmonellae may contaminate chelonian eggs during cloacal passage or during nest deposition in moist soil or sand.[1] *Salmonella* spp. can invade the shell within 1 hour of exposure and ovarial colonization, transovarial transmission, and transmission during pregnancy in ophidians can occur.[1] Salmonellae may be present inside the eggs, as some ophidian species lay eggs surrounded by a thin membrane, instead of a true shell, which may be easily invaded by microorganisms. Leptospirosis also has been associated with people handling and collecting eggs from crocodilians.[1]

Vertical transmission of viral disease in reptiles has received little study. PCR, viral culture, and histopathology of the fetus may provide clues to virus-related morbidity and mortality in reptiles and amphibians. Paramyxovirus grows readily in cobra *eggs*, and Western equine encephalitis virus has been vertically transmitted between dam and offspring in the garter snake without reported health concerns for the offspring or dam.[10]

Reptile eggs may tolerate fungal growth on shells and shell membranes of developing eggs based on an albumen with antimicrobial properties;[5] however, *Fusarium solani* has been isolated from crocodile nesting material, shell, and dead-in-shell crocodile embryos. Hatchling snakes may be able to abandon the yolk sac when incubated at low temperature or based on fungal infection.[5] In clutches of the Iberian rock lizard, *Lacerta monticola,* the hypotheses that "Infertile and non-viable fertile eggs within a reptile clutch may decrease the incubation success of the remaining eggs, as (1) opportunistic pathogens may use the nutrient resources provided by dead eggs to colonize the clutch and spread to and kill viable eggs; and (2) odors released by spoilt eggs may attract predators to the clutch."[10] were evaluated in a study comparing incubation success of clutches of solely of fertile eggs and clutches containing some dead eggs. In the lab, fungi (*Fusarium* sp. and *Gliocladium* sp.) colonized both infertile eggs and fertile eggs that died during incubation and spread to and killed adjacent eggs. Hatchlings from fungal eggs hatched earlier, before absorbing the whole yolk and were smaller and lighter than nonfungal hatchlings; however, a field experiment did not corroborate the fungal pathogenic effects observed in the lab nor did they confirm that clutches containing dead eggs suffer higher predation.[11] Consider removal of eggs that fail to develop signs of embryonic growth (white patch/chalking) to another incubator as growth media for bacteria and fungus.

Cleaning adherent dirt and vegetation from the rigid turtle or crocodilian shell with tap water also may be acceptable prior to incubation.[5] Experimentally, partial hypoxia of the reptile shell affects embryos but is not associated with reduced hatch rate so removal of adherent material, unless suspected associated with a heavy microbe load is likely unnecessary.[12] The red-banded snake (*Lycodon rufozonatum*) with a parchment egg shell and Chinese soft-shelled turtle (*Pelodiscus sinensis*) with rigid egg shell eggs were obtained from nests, and 50% covered with non-toxic and oxygen-impermeable beeswax on the upper half and bottom half of the eggs. In this controlled study the growth pattern of the opaque white patch in turtle eggs, embryonic heart rate, developmental rate and hatchling traits were measured for the eggs. Snake embryos increased their heart rate while turtle embryos increased growth of the opaque white patch. Incubation period and hatching success of embryos were not affected by the hypoxia treatment; however, embryos from the upper-half hypoxia treatment used less energy from yolk and were smaller hatchlings.

Although egg treatment is generally not recommended based on the porous nature of reptile shells, forest dragon *Hypsilurus spinipes* and green turtle eggs removed from dead females dipped in a 1% solution of betadyne experienced no apparent embryonic harm.[5] Sodium hypochlorite and polyhexamethylene biguanide have been used as an egg wash to reduce *Salmonella* spp. on red-eared slider turtle (*Trachemys scripta elegans*) egg surfaces and hatchlings.[13,14] Many factors affect hatchling traits in oviparous reptiles (Table 5).

**Table 5.** Oviparous reptile diversity regarding effect of incubation factors and hatchling traits.

| | Incubation Factors | Hatching traits | Exemplar species |
|---|---|---|---|
| Squamates without parental care | Mean temp<br>Diel temp<br>Temp variance<br>Moisture<br>Gradual seasonal clines<br>Diel temp range<br>Timing of hot/cold weather | Sex, size, hatching time, % hatch, speed, shape, head size, behavior, growth, deformities, egg mortality, residual yolk, antipredator tactics, energy resources | *Eu macularius*<br>*Pituophis melanoleucas* |
| Squamates with parental care | Mean temp, moisture, thermal variance, basking time | Birthdate, size, speed, sex, body shape, scalation percent, % still born, activity level, locomotor speed, hatch %, survival, residual yolk | *Python regius* |

## References


1. Ebani VV. *Asian Pacific Journal of Tropical Medicine.* 2017;10(8):723-728.

2. Packard GC, Tracy CR, Roth JJ. The physiological ecology of reptilian eggs and embryos and the evolution of viviparity within the class Reptilia. *Biol Rev.* 1977;52(1):71-105.

3. Wiemann J, Yang TR, Norell MA. Dinosaur egg colour had a single evolutionary origin. *Nature.* 2018;563(7732):555.

4. Wiemann J, Yang TR, Sander PN, et al. Dinosaur origin of egg color: oviraptors laid blue-green eggs. *Peer J.* 2017;29:5:e3706.

5. Deeming DC. *Reptilian Incubation: Environment, Evolution and Behaviour.* Nottingham, UK: Nottingham University Press;2004.

6. Rafferty AR, Reina RD. Arrested embryonic development: a review of strategies to delay hatching in egg-laying reptiles. *Proc Royal Soc B: Biol Sci.* 2012;21;279(1737):2299-308.

7. Walls JG. Candling reptile eggs. Reptiles. Available at: http://www.reptilesmagazine.com/Breeding-Lizards/Candling-Eggs/ Accessed May 3, 2019.

8. Fogel D. Red-footed tortoise breeding. Reptiles. Available at: http://www.reptilesmagazine.com/Red-Footed-Tortoise-Breeding/ Accessed May 3, 2019.

9. Spencer RJ, Thompson MB, Banks PB. Hatch or wait? A dilemma in reptilian incubation. *Oikos.* 2001;93(3):401-406.

10. Ariel E. Viruses in reptiles. *Vet Res.* 2011;42(1):100.

11. Moreira PL, Barata M. Egg mortality and early embryo hatching caused by fungal infection of Iberian rock lizard (*Lacerta monticola*) clutches. *Herpetol J.* 2005 Oct 1;15(4):265-72.

12. Tang WQ, Zhao B, Chen Y, Du WG. Reduced egg shell permeability affects embryonic development and hatchling traits in *Lycodon rufozonatum* and *Pelodiscus sinensis*. *Integr Zool.* 2018;13:58-69.

13. Mitchell MA, Adamson TW, Singleton CB, et al. Evaluation of a combination of sodium hypochlorite and polyhexamethylene biguanide as an egg wash for red-eared slider turtles (*Trachemys scripta elegans*) to suppress or eliminate *Salmonella* organisms on egg surfaces and in hatchlings. *Am J Vet Res.* 2007;68(2): 158-64.

14. Mitchell TS, Janzen FJ. Substrate influences turtle nest temperature, incubation period, and offspring sex ratio in the field. *Herpetologica.* 2019;75(1):57-62.

## Additional Reading

- Ballen CJ, Shine R, Olsson MM. Developmental plasticity in an unusual animal: the effects of incubation temperature on behavior in chameleons. *Behaviour.* 2015;152(10):1307-1324.
- Birchard GF, Marcellini D. Incubation time in reptilian eggs. *J Zool.* 1996;240(4):621-635.
- Bodensteiner BL, Mitchell TS, Strickland JT, Janzen FJ. Hydric conditions during incubation influence phenotypes of neonatal reptiles in the field. *Funct Ecol.* 2015;29(5):710-717.
- Booth DT. Influence of incubation temperature on hatchling phenotype in reptiles. *Physiol Biochem Zool.* 2006;79(2):274-281.
- Bowden RM, Carter AW, Paitz RT. Constancy in an inconstant world: moving beyond constant temperatures in the study of reptilian incubation. *Integr Comp Biol.* 2014;54(5):830-840.
- Brusch IV GA, DeNardo DF. Egg desiccation leads to dehydration and enhanced innate immunity in python embryos. *Dev Comp Immunol.* 2019;90:147-151.
- Burger J. Effects of incubation temperature on behavior of hatchling pine snakes: implications for reptilian distribution. *Behav Ecol Sociobiol.* 1991;28(4):297-303.
- Bustard HR. Tail abnormalities in reptiles resulting from high temperature egg incubation. *Br J Herpetol.* 1969b;4:121-123.
- Cagle KD, Packard GC, Miller K, Packard MJ. Effects of microclimate in natural nests on development of embryonic painted turtles, *Chrysemys picta*. *Funct Ecol.* 1993;7:653-660.
- Carter AW, Bowden RM, Paitz RT. Seasonal shifts in sex ratios are mediated by maternal effects and fluctuating incubation temperatures. *Funct Ecol.* 2017;31(4):876-884.
- Dayananda B, Webb JK. Incubation under climate warming affects learning ability and survival in hatchling lizards. *Biol Lett.* 2017;13(3):20170002.

- Du WG, Ji X. The effects of incubation thermal environments on size, locomotor performance and early growth of hatchling soft-shelled turtles, *Pelodiscus sinensis*. *J Thermal Biol.* 2003;28(4):279-286.
- Du WG, Radder RS, Sun B, Shine R. Determinants of incubation period: do reptilian embryos hatch after a fixed total number of heart beats? *J Exp Biol.* 2009;212(9):1302-1306.
- Keller KA. Reptile perinatology. *Vet Clin North Am Exot Anim Pract.* 2017;20(2):439-454.
- Marco D, Thompson MB. Embryonic development times and egg retention in four species of sceloporine lizards. *Funct Ecol.* 1992;6:436-444.
- McGlashan JK, Thompson MB, Van Dyke JU, Spencer RJ. Thyroid hormones reduce incubation period without developmental or metabolic costs in Murray River short-necked turtles (*Emydura macquarii*). *Physiol Biochem Zool.* 2017;90(1):34-46.
- Noro M, Uejima A, Abe G, et al. Normal developmental stages of the Madagascar ground gecko Paroedura pictus with special reference to limb morphogenesis. *Dev Dyn.* 2009;238(1):100-109.
- Outerbridge ME, O'Riordan R, Fort DJ, Davenport J. Ecotoxicological assessment of diamondback terrapin (*Malaclemys terrapin*) pond habitat, prey and eggs in Bermuda. *Marine Pollut Bull.* 2016;15;102(1):36-43.
- Portas TJ. Reproductive strategies in reptiles. In: Doneley B, Monks D, Johnson R, Carmel B, eds. *Reptile Medicine and Surgery in Clinical Practice.* Ames, IA: Wiley-Blackwell. 2018:91.
- Siviter H, Deeming DC, Rosenberger J, et al. The impact of egg incubation temperature on the personality of oviparous reptiles. *Anim Cogn.* 2017;20(1):109-116.

# Pathology Lesions of Reptile, Amphibian, and Fish Eggs and Embryos

Drury R. Reavill, DVM, DABVP (Avian Practice), DABVP (Reptile and Amphibian Practice), DACVP, and
David J. Gasper, DVM, PhD, DACVP

**Session #2845**

*Affiliation:* From Zoo /Exotic Pathology Service, 6020 Rutland Drive #14, Carmichael, CA 95608-0515, USA (Reavill), Pacific Zoo & Wildlife Diagnostics, San Diego, CA 92027, USA (Gasper).

*Abstract:* This review focuses on egg and embryo pathology as well as early neonatal development. This information should assist in determining cause of reproductive failures and neonatal death and lesions that adversely affect successful breeding of these species. This is one part (pathology lesions) of a three-part ABVP-level review of the anatomy, clinical approach to egg failure, and embryo and neonatal diseases of reptiles, as well as some amphibian and fish examples.

## Infectious Disease Agents

Agamid atadenovirus 1 infection has become established in U. S. lizard breeding groups of bearded dragons (*Pogona* species) and has been identified around the world including in their country of origin, Australia.[1] The common clinical signs in juvenile dragons are a progressive weakness, neurological signs, and anorexia. It is speculated that there is possible vertical transmission through the egg in utero or at time of oviposition. Histologically the lesions of severe acute coagulative hepatocellular necrosis and the large intranuclear basophilic inclusion bodies in the hepatocytes and enterocytes are diagnostic for the adenovirus infection.

Ranavirus in amphibians, larvae and adults, can be a persistent infection within some environments. It is transmitted by direct and indirect routes; contaminated environment, direct contact with infected individuals, and ingestion by predation and/or cannibalism. Outbreaks of the infection in tadpoles are suspected to be due to stressors or infections in naïve hosts. Grossly the tadpoles will develop edema, erythema, enlarged livers, and hemorrhage.[2]

*Saprolegnia*, *Basidiobolus*, and *Mucor* spp. are ubiquitous aquatic molds that opportunistically infect local skin defects or contaminate egg masses. Contaminated eggs should be removed and animals can be treated with salt, malachite green, or copper sulfate baths.[3]

*Fusarium solani* is responsible both for superficial and invasive infection in sea turtle nest contamination, causing massive losses during hatching. Loggerhead turtles (*Caretta caretta*) nests were contaminated with this fungus and this resulted in skin and internal infections of the hatchlings.[4]

Severe Perkinsea infection (SPI) is an emerging disease of frogs responsible for mass mortalities of 10 tadpole species across the United States. The affected amphibians are from boreal to subtropical areas. The protozoa belongs to the phylum Perkinsozoa that form a distinct group referred to as the Pathogenic Perkinsea Clade of frogs. Other Perkinsea species within the genus Perkinsus (eg, *P. marinus, P. olseni*) are responsible for mass mortalities and severe economic losses of mollusks. The infection results in lesions of the tadpoles and adult frogs have been diagnosed. The tadpoles have coelomic distention, erythema, petechial, hepatomegaly, splenomegaly, enlarged kidneys, and segmental thickening of the intestinal wall. Histologic lesions are associated with invasion of numerous tissues by myriads of intracellular and extracellular Perkinsea- like organisms. The most extensively affected organs are liver, spleen, kidney and pancreas.[5,6]

Tetrahymena-like ciliates are associated with mortality of the Northern corroboree frog (*Pseudophryne pengilleyi*) tadpoles and eggs, a species found in Australia and New Zealand. The primary lesions in the tadpoles are inflammation and necrosis of the dermis and skeletal muscle. Some of the egg capsules also contained ciliates, but most were overgrown with bacteria and fungi. The eggs of this frog are eggs are laid out of water in a terrestrial nest. Nests of eggs require submersion in water, to stimulate hatching, after which the tadpoles make their way to alpine pools. The adults in this outbreak, were not affected. The affected tadpoles were found dead or behaving abnormally with minimally raised cutaneous pale tan foci. Affected eggs initially appeared fertilized, but had arrested development.[7]

Aquatic snails release infective cercariae of the trematode *Ribeiroia ondatrae* that encyst as metacercariae in the limb buds of juvenile amphibians. Infection impairs limb development resulting in missing and extra limbs or digits, with the frequency of malformation positively correlated with infection intensity.[8] Similar lesions were produced in *Hyla savignyi* and *Pelophylax bedriagae* by an echinostomatid trematode. The tadpoles were malformed and edematous.[9]

Myxosporea are a metazoan parasite closely related to jellyfish. The intermediate hosts are primarily aquatic vertebrates fish, reptiles, and amphibians. Some of the known definitive hosts are annelids. Infections can range from inapparent to severe. *Myxobolus cerebralis* has been associated with significant morbidity and mortality in captive raised and wild fish. *Cystodiscus axonis* and *C. australis* are pathogens from many native and exotic Australian frogs. *C. axonis* plasmodia in the brains of tadpoles and adults results hemorrhage, necrosis, and a vasculitis, with associated neurological behavioral changes and death. Both species develop in the bile dusts and tadpoles resulting in damage to the liver interfering with metamorphosis and immune function.[10]

The trematode *Telogaster opisthorchis* caused severe malformations in *Galaxias anomalus*, a threatened New Zealand freshwater fish. The heavy infections by the metacercarial cysts of the trematode resulted in severe spinal and fin malformations of juvenile fish. The trematode metacercariae physically impair developing tissue.[11]

## Toxins

Environmental contaminants have long been known to be associated with reproductive and congenital defects in reptiles, amphibians, and fish. These effects have ranged from thyroid dysfunction, hatching success, and gross birth deformities.[12] There are many studies measuring and/or exposing eggs and embryos to toxins to specifically understand the mechanisms. Lab fish and amphibians (usually *Xenopus laevis*) are very common species used for these studies. Commercially important fish species are also frequently evaluated in light of a food source for humans.

Organochlorinated pesticides have been identified in several reptile, amphibian, and fish species and associated with a variety of reproductive lesions. These pesticides were used extensively from the 1940s through the 1960s in agriculture and mosquito control. Representative compounds in this group include DDT, methoxychlor, dieldrin, chlordane, toxaphene, mirex, kepone, lindane, and benzene hexachloride. These persist in the environment and continued to have potential adverse effects on wildlife and human health. Many of these contaminants bioconcentrate in the lipid rich yolk. The chemicals deposited in terrestrial eggs persist throughout development of the embryo and can be transferred through the eggshell.[12]

In the American alligator (*Alligator mississippiensis*) increased histologic lesions were present in hatchlings with increasing elevations of organochlorinated pesticides in the yolk. Delayed growth and pneumonia were considered secondary to exposure.[13] Pesticide spills have been associated with abnormal gonads, altered hormone concentrations and reduced penis size.[12]

*Xenopus laevis* embryos have been used for many toxicity studies. Carbaryl, a frequently used insecticide in the Carbamate chemical family because of its low toxicity and short lifetime in the environment, results in musculoskeletal lesions of the developing frog larvae. Its teratogenic effects present as an abnormal tail flexure. The skeletal muscle lesions consist of reduced size of the myotomes, disorganized myocytes, and wavy or bent notochords. A similar lesion is described in toxic exposure to chlorpyrifos (an organophosphorus insecticide) in larval *Rana dalmatina*.[14] This axial–skeletal damage is hypothesized to be related to the inhibition of acetylcholinesterase.[15] *Xenopus* larvae are also used in pharmaceuticals studies especially to evaluate hazards of contaminated surface water. Tails flexures, abnormal gut coiling, and thoracic edema were some of the teratogenic effects to various commonly used pharmaceuticals.[16]

In areas of pesticide runoffs (some of which have estrogenic activity), the wild populations of frogs and toads have developmental abnormalities prominent at metamorphosis. These include ectromelia and ectrodactyly. Feminization of male toads have resulted in changes to the forelimb lengths, coloration changes and reduced plasma testosterone concentrations.[12,17] Elevations in water levels of boron (boron compounds such as borax and boric acid are used as fertilizers in agriculture) and nitrates resulted in frogs and salamanders the hatchlings had a curling defect where the larvae had difficulties in swimming. Other salamanders had altered gill morphology.[18]

Zinc concentrations in the water as low as 0.2 mg/L can result in severe gastrointestinal and skin lesions in adult newts and as well as in developing juvenile amphibians. High water phosphate concentrations can interfere with calcium metabolism contributing to metabolic bone disease in developing frog larvae.[19]

Elevated nitrate levels in surface water (from fertilizers, manure, and lime applications) alters thyroid function in fish and results in reduced embryo weights and productivity.[12]

Brevetoxins, a neurotoxic lipophilic polyether toxin, are produced by the red tide dinoflagellate, *Gymnodinium breve*. Studies using Medaka fish (*Oryzias latipes*) have demonstrated teratogenic effects. Exposure resulted in lateral curvature of the spinal column, herniation of brain meninges through defects in the skull, and malpositioned eye. Head first hatching abnormalities were also commonly observed.[20]

There is increased egg and larvae increased mortality in gilthead seabream (*Sparus aurata*) with acute copper toxicity. In subacute exposure, morphological alterations or abnormalities in embryos included irregular shapes of chorion, opacity and vitellus retraction/degeneration. In larvae there was poor capacity to swim, trembling, myoskeletal defects, opacity and exophthalmia.[21]

## Nutrition

A study was conducted to compare thiamine concentrations in American alligators (*Alligator mississippiensis*) and Florida largemouth bass (*Micropterus salmoides floridanus*). Results suggest that thiamine deficiency might be playing an important role in alligator embryo survival but not in reproductive failure and recruitment of largemouth bass. The cause(s) of this thiamine deficiency are unknown but might be related to differences in the nutritional value of prey items across the sites studied and/or to the presence of high concentration of contaminants in eggs.[22,23]

Bioaccumulated selenium in the eggs and larvae of rainbow trout (*Oncorhynchus mykiss*) was associated with developmental abnormalities, specifically craniofacial defects, skeletal deformities, and edema.[24]

Low dietary availability of phosphorus in Atlantic salmon (*Salmo salar L*) in first feeding fry resulted in spinal deformities. This appeared radiographically as morphologic abnormal vertebra, bone density, lordosis, and kyphosis.[25]

## Environmental Factors

Some studies have reviewed normal development of amphibians while in space. Microgravity environments (space) have resulted in lesions of the lungs and tails for urodele and anuran species. Tadpoles develop lordosis and do not adequately inflate their lungs. The lung collapse is due to the animals not identifying the air-water interface.[26]

Both UV-B and aquatic hypoxia have negative effects on developing embryos and larvae of the striped marsh frog (*Limnodynastes peronei*). Exposure of embryonic and larval amphibians to UV-B radiation has been shown to cause increased mortality rates, reduced growth and development rates, delayed metamorphosis, reduced locomotor performance, developmental abnormalities, behavioral changes, and increased susceptibility to disease. This is suspected to be due to DNA damage from the UV-B exposure. Low dissolved oxygen levels can delay embryonic development, as well as result in behavioral changes such as increased activity and increased respiration. Naturally hypoxic waters often occur as the result of temperature changes, stratification, and dense population of aquatic organisms.[27]

Water temperature extremes have been associated with embryo morphological defects in fish. Elevated water temperature (25°C) resulted in a shorter incubation time for the development of the annual fish, *Cynopoecilus melanotaenia*, and increased defects.[28]

## Developmental Abnormalities of Unknown Etiology

Retained yolk sacs are more commonly described in birds and are associated with considerable mortality and morbidity. It has been described in a Burmese python (*Python bivittatus*).[29] The yolk sac is normally withdrawn into the body cavity prior to hatching. The contents of lipid and proteins will be absorbed by a yolk stalk into the intestine post-hatching. The python presented with coelomic distention at the location of the umbilicus at 9 days post-hatch. The mass could be identified on radiographs. The persistent yolk sac was successfully surgically removed. The mass was of yolk material, macrophage proliferation, and a few bacteria.

Muscular dystrophy is described in variety of lizards and snakes. The lesion is characterized by irregular atrophy, with myofibers lost and replaced by fat. In a series of evaluations, no sex bias or species-specific findings were noted.[30]

Schistosomus reflexus (SR) is a rare and fatal congenital malformation syndrome occurring during embryonic development. These congenital malformations have been described in the olive ridley sea turtle (*Lepidochelys olivacea*). Malformed turtles showed a marked lordosis and other vertebral column malformations such as scoliosis, kyphosis, and vertebral and rib alterations (number, shape, and fusion). The shell was underdeveloped: bones and scutes agenesis, subnumerary and deformed scutes. Flippers displayed malformations: arthrogryposis, and dysmelias: amelia, phocomelia, syndactily, among others. The high prevalence of SR in this species suggests a primary genetic etiology, although environmental factors may be involved.[31]

There are surveys of congenital malformations of sea turtles (hawksbills, green sea turtles, and olive ridleys). These include: leucism, conjoined twins, anasarca, dwarfism, gigantism, bicephaly, acephaly, microcephaly, hydrocephaly, encephalocele, skull dysostosis, anencephaly, exencephaly, eyelid hyperplasia, microphthalmia, cyclopia, anophthalmia, rhinocephaly, laterognathia, brachygnathia, prognathia, agnathia, cervical cysts, neck pseudofistula, compressed carapace, misshapen scutes, supernumerary scutes, kyphosis, scoliosis, osteoagenesis, plastron metaplasia, anury, brachyury, cloacal atresia, micromelia, aplastic claws, macrodactyly, polydactyly, diplopodia, hypoplasia, phocomelia, and amelia.[32]

One detailed case of unilateral pulmonary aplasia, asymmetrical carapacial kyphosis, and mild scoliosis was described in an adult common snapping turtle (*Chelydra serpentina*). In this animal there was some aquatic locomotion and buoyancy difficulties. The congenital defect was non-fatal and suspected to be a genetic mutation.[33]

Cardiac malformation in two juvenile ball pythons (*Python regius*) resulted in equal systolic pressures in the pulmonary and systemic arteries. The defect was of the reduced size of the muscular ridge of the ventriculus.[34]

## Dystocia

Dystocia and subsequent surgical intervention has been described in Rio Cauca caecilians or rubber eel (*Typhlonectes natans*) a viviparous amphibian. Briefly the embryo development is divided into 34 stages; intrauterine hatching occurs between stages 25 and 26, metamorphosis from stage 30 to 32, and birth at stage 34. Anorexia in the female and increasing time between passing of live and some dead young triggered the decision for cesarean surgery. Specific causes for the dystocia was not identified.[35]

## References


1. Doneley RJT, Buckle KN, Hulse L. Adenoviral infection in a collection of juvenile inland bearded dragons (*Pogona vitticeps*). *Aust Vet J.* 2014;92(1-2):41-45.

2. Gray MJ, Miller DL, Hoverman JT. Ecology and pathology of amphibian ranaviruses. *Dis Aquat Org.* 2009;87:243-266.

3. Baitchman EJ. Amphibian medicine. *Proc Annu Conf Am Assoc Zoo Vet.* 2008;45-50.

4. Garcia-Hartmann M, Hennequin C, Catteau S, Béatini C, Blanc V. Clusters of *Fusarium solani* infection in juvenile captive born *Caretta caretta* sea turtles. *J Mycologie Médicale.* 2017;27:113-118.

5. Isidoro-Ayza M, Grear DA, Chambouvet A. Pathology and case definition of severe Perkinsea infection of frogs. *Vet Pathol.* 2019;56(1):133-142.

6. Landsberg JH, Kiryu Y, Tabuchi M, et al. Co-infection by alveolate parasites and frog virus 3-like ranavirus during an amphibian larval mortality even in Florida, USA. *Dis Aquat Org.* 2013;105:89-99.

7. Holz PH, Portas T, Donahoe S, et al. Mortality in northern corroboree frog tadpoles (*Pseudophryne pengilleyi*) associated with Tetrahymena-like infection. *Aust Vet J.* 2015;93(8):295-297.

8. Johnson PTJ, Lunde KB, Ritchie EG, Launer AE. The effect of trematode infection on amphibian limb development and survivorship. *Science.* 1999;284:802-804.

9. Goren L, Routtu J, Ben-Ami F. Trematode-associated morbidity and mortality of tadpoles in Israel. *Parasitol Res.* 2014;113(10):3833-3841.

10. Hartigan A, Dhand NK, Rose K, et al. Comparative pathology and ecological implications of two myxosporean parasites in native Australian frogs and the invasive cane toad. *PLoS One.* 2012;7(10):e43780.

11. Kelly DW, Thomas H, Thieltges DW, et al. Trematode infection causes malformations and population effects in a declining New Zealand fish. *J Anim Ecol.* 2010;79(2):445-452.

12. Hamlin HJ, Guillette Jr LJ. Birth Defects in Wildlife: The Role of Environmental Contaminants as Inducers of Reproductive and Developmental Dysfunction. Syst Biol Reprod Med. 2010;56(2):113-121.

13. Sepulveda MS, Del Piero F, Wiebe JJ, et al. Necropsy findings in American alligator late-stage embryos and hatchlings from northcentral Florida lakes contaminated with organochlorine pesticides. *J Wildl Dis.* 2006;42(1):56-73.

14. Bernabò I, Sperone E, Tripepi S, Brunelli E. Toxicity of chlorpyrifos to larval *Rana dalmatina*: acute and chronic effects on survival, development, growth and gill apparatus. *Arch Environ Contam Toxicol.* 2011;61(4):704-718.

15. Bacchetta R, Mantecca P, Andrioletti M, et al. Axial-skeletal defects caused by Carbaryl in *Xenopus laevis* embryos. *Sci Total Environ.* 2008;392(1):110-118.

16. Richards SM, Cole SE. A toxicity and hazard assessment of fourteen pharmaceuticals to *Xenopus laevis* larvae. *Ecotoxicology.* 2006;15(8):647-656.

17. Nishimura N, Fukazawa Y, Uchiyama H, Iguchi T. Effects of estrogenic hormones on early development of *Xenopus laevis*. *J Exp Zool.* 1997;278(4):221-233.

18. Laposata MM, Dunson WA. Effects of boron and nitrate on hatching success of amphibian eggs. *Arch Environ Contam Toxicol.* 1998;35(4):615-619.

19. Geschke K, Goad T, Lambert C, et al. Veterinary challenges of raising threatened orange-bellied frogs (*Geocrinia vitellina*). *Proc Annu Conf Assoc Avian Vet Austr Comm, Unusual Pet Avian Vet.* 2014.

20. Kimm-Brinson KL, Ramsdell JS. The red tide toxin, brevetoxin, induces embryo toxicity and developmental abnormalities. *Environ Health Perspect.* 2001;109(4):377-381.

21. Oliva M, Garrido MDC, Pérez E, de Canales MLG. Evaluation of acute copper toxicity during early life stages of gilthead seabream, *Sparus aurata*. *J Environ Sci Health A Tox Hazard Subst Environ Eng.* 2007;42(4):525-533.

22. Sepulveda MS, Wiebe JJ, Honeyfield DC, Rauschenberger HR, Hinterkopf JP, Johnson WE, Gross TS. Organochlorine pesticides and thiamine in eggs of largemouth bass and American alligators and their relationship with early life-stage mortality. *J Wildl Dis.* 2004;40(4):782-786.

23. Rauschenberger RH, Sepúlveda MS, Wiebe JJ, et al. Nutrient and organochlorine pesticide concentrations in American alligator eggs and their associations with clutch viability. *J Aquat Anim Health.* 2009;21(4):249-261.

24. Holm J, Palace V, Siwik P, et al. Developmental effects of bioaccumulated selenium in eggs and larvae of two salmonid species. *Environ Toxicol Chem.* 2005;24(9):2373-2381.

25. Sullivan M, Reid SWJ, Ternent H, et al. The aetiology of spinal deformity in Atlantic salmon, Salmo salar L.: influence of different commercial diets on the incidence and severity of the preclinical condition in salmon parr under two contrasting husbandry regimes. *J Fish Dis.* 2007;30(12):759-767.

26. Wassersug RJ, Yamashita M. The mechanics of air-breathing in anuran larvae: implications to the development of amphibians in microgravity. *Adv Space Res.* 2000;25(10):2007-2013.

27. Bernal MH, Alton LA, Cramp RL, Franklin CE. Does simultaneous UV-B exposure enhance the lethal and sub-lethal effects of aquatic hypoxia on developing anuran embryos and larvae? *J Comp Physiol B.* 2011;181(7):973-980.

28. Arenzon A, Lemos CA, Bohrer MBC. The influence of temperature on the embryonic development of the annual fish Cynopoecilus melanotaenia (*Cyprinodontiformes, rivulidae*). *Braz J Biol.* 2002;62(4B):743-747.

29. Bodri MS, Hendrick MJ, O'Brien RT, Sadanaga KK. Retained caseous yolk sac in a Burmese python (*Python molurus bivittatus*). *J Wildl Dis.* 1990;26(4):564-566.

30. Stolk A. Muscular dystrophy in fishes, amphibians and reptiles. *Acta Morphol Neerl Scand.* 1962;5(0): 117-139.

31. Bárcenas-Ibarra A, Rojas-Lleonart I, García-Gasca A. Congenital malformations in sea turtles: first report of schistosomus reflexus syndrome. *34th Annu Symp Sea Turtle Biol Conserv.* 2014.

32. Bárcenas-Ibarra A, de la Cueva H, Rojas-Lleonart I, et al. First approximation to congenital malformation rates in embryos and hatchlings of sea turtles. *Birth Defects Res A Clin Mol Teratol.* 2015;103(3):203-224.

33. Schachner ER, Sedlmayr JC, Schott R, et al. Pulmonary anatomy and a case of unilateral aplasia in a common snapping turtle (*Chelydra serpentina*): developmental perspectives on cryptodiran lungs. *J Anat.* 2017;231(6):835-848.

34. Jensen B, Wang T. Hemodynamic consequences of cardiac malformations in two juvenile ball pythons (*Python regius*). *J Zoo Wildl Med.* 2009;40(4):752-756.

35. Barbon AR, Goetz M, López J, Routh A. Uterine Rupture And Cesarean Surgery In Three Rio Cauca Caecilians (*Typhlonectes Natans*). *J Zoo Wildl Med.* 2017;48(1):164-170.

**Gap in pagination due to formatting issues.**

**Page 649**

# Dietary Cause of Gastroduodenal Perforations in Spiny Softshell Turtles (*Apalone spinifera*)

Shannon T. Ferrell, DVM, DABVP (Avian Practice), Dipl ACZM,
Isabelle Devost, MSc,
Lois Lazure, MSc,
Émilie L. Couture, DMV, IPSAV

**Session #2725**

*Affiliation:* From the Zoo de Granby, 525 rue Saint-Hubert, Granby, QC J2G 5P3 Canada.

As part of the recovery plan for the endangered spiny softshell turtle (*Apalone spinifera*) in the province of Québec (Canada), a headstarting program was implemented with the objective of increasing juvenile survival over the first winter. High mortality rates were observed during the first three years of the program. An unusually high incidence of gastroduodenal perforations and resulting coelomitis was linked to these mortalities. Microscopically, the coelomitis consisted of serosal mesothesial cell hypertrophy, fibrinous to granulomatous inflammation, visceral adhesions, as well as the presence of bacterial colonies, and alimentary particles. Over the years, measures were taken such as the removal of sun-dried red shrimps and freeze-dried *Gammarus* shrimps from the diet as well as the strict control of offered food quantities; however, the elimination of these potential risk factors did not prevent mortalities due to perforations. In order to investigate the impact of diet on the occurrence of gastrointestinal perforations, half of the headstart individuals (n = 11) was fed with a dried, pelleted, turtle food (soaked in water before being offered), and the other half (n = 11) was fed with a soft, moistened, gel mixture. All of the mortalities (n = 4) occurred in the group fed with dried turtle pellet, and three of these deaths were linked to gastrointestinal perforations. Moreover, higher growth rates were observed in the gel mixture group. The results of this study suggest that the replacement of the dried pelleted food by a gel mixture is appropriate to prevent gastrointestinal perforations and to promote growth in this species.

*Acknowledgments:* The authors thank Scott Gillingwater for his dietary assistance and the Zoo de Granby for its support and assistance in the care of these animals.

# Effect of Bone Morphogenetic Protein on Chelonian Experimental Carapace Defects

Charles Innis, VMD, DABVP (Reptile and Amphibian Practice),
Christopher Wilson, PhD,
Howard Seeherman, PhD, VMD,
Brian Stacy, DVM, DACVP

**Session #2654**

*Affiliation:* From New England Aquarium, Central Wharf, Boston, MA 02110, USA (Innis), Bioventus, St. Mary's St., Boston, MA 02115, USA (Wilson, Seeherman), University of Florida College of Veterinary Medicine, SW 35th Drive, Gainesville, FL 32608, USA (Stacy).

*Note:* Abstract format selected by author(s). See Front Matter for description of conference proceedings manuscripts submitted as scientific abstracts.

*Abstract*: Anecdotal reports suggest that bone morphogenetic protein (BMP) may be useful for treating chelonian shell defects, but controlled studies are lacking. Under general anesthesia, 2 cm diameter, full-thickness carapace defects were created bilaterally in the second costal scute of twelve red-eared slider turtles (*Trachemys scripta elegans*). Defects were managed for ten days using negative pressure wound therapy, followed by bandages. Three weeks post-operatively, one defect on each turtle was injected with 0.1 mg recombinant human BMP-2 (Bioventus, LLC, Durham, NC, USA) carried in bovine type 1 collagen (Sigma Aldrich, St. Louis, MO, USA), while the contralateral defect was injected with only collagen. Four turtles were euthanized at each of three time points (2, 4, 12 weeks post-treatment). Defects were evaluated by gross appearance, computed tomography, and histology. No difference was noted between the treated and untreated defects. Nonetheless, histologic assessment provided novel characterization of the progression of chelonian shell healing under controlled conditions. Examination indicated that margins of exposed carapace bone underwent osteonecrosis. Re-epithelialization and osteoclastic activity undermined necrotic bone, which was eventually sloughed. Inflammation of the deep wound bed, including the peri-coelomic membrane, resolved with completion of re-epithelialization and was associated with the transition of granulation tissue into mature fibrous connective tissue. Osteoid and woven bone formed along remodeled margins and began to bridge the wound along its deep surface within this fibrous connective tissue. The surgical model described here may be useful for future studies of chelonian shell repair and healing, including studies that evaluate alternative formulations of BMP.

# Plasma Concentrations of Oxytetracycline after Multiple Subcutaneous Injections in Three Sea Turtle Species

Charles Innis, VMD, DABVP (Reptilian and Amphibian Practice),
Adam Kennedy, BS,
Julika Wocial, MS,
Elizabeth Burgess, PhD,
Mark Papich, DVM, MS, DACVCP

**Session #2657**

*Affiliation:* From New England Aquarium, Central Wharf, Boston, MA 02110, USA (Innis, Kennedy, Wocial, Burgess), North Carolina State University, College of Veterinary Medicine, William Moore Drive, Raleigh, NC 27607, USA (Papich).

*Note:* Abstract format selected by author(s). See Front Matter for description of conference proceedings manuscripts submitted as scientific abstracts.

*Abstract*: Oxytetracycline (42 mg/kg SC q6d; Bio-Mycin 200, Boehringer Ingelheim, St. Joseph, MO, USA) was injected into the dorsal shoulder of 29 sea turtles during clinical management after stranding, including 10 Kemp's ridley (*Lepidochelys kempii*), 10 loggerhead (*Caretta caretta*), and 9 green turtles (*Chelonia mydas*). Before injection, drug was diluted into lactated ringer's solution (5 ml/kg) or into fluid chosen specifically for an individual turtle by the attending veterinarian (eg, reptile ringer's solution, ranging from 5 to 20 ml/kg). Turtles were treated for as long as clinically indicated prior to discontinuing the drug (range 2-10 doses, median 4). After the final injection, plasma was harvested at 1, 2, 4, 8, 72, 144, and 288 hours after injection for assessment of plasma oxytetracycline concentration. To limit the volume of blood collected, samples for only 3 or 4 time points were collected from any individual. No significant difference in pharmacokinetic parameters were detected between species. Population pharmacokinetics using nonlinear mixed effects modeling indicated half-life = 36 hours, volume of distribution = 0.78 l/kg, and area-under-the-curve = 2799 μg×h/ml. Trough concentrations at 288 hours post-injection were undetectable for the majority of individuals (< 0.05 μg/ml). Results indicate that this delivery route, dose, and dosing interval may provide antibiotic efficacy for approximately six days for organisms with minimum inhibitory concentrations of 4 μg/ml or less.

# Disease Processes Associated with Propagation for Reintroduction of Threatened Sierra Nevada Frogs

Michael M. Garner, DVM, DACVP,
Adrian Mutlow, MA, VetMB, MSc, MRCVS,
Karen Emanualson,
Julie Barnes,
Kelly Garner,
Cherie Briggs,
Manuel Garcia,
Shannon Nodolf

**Session #2681**

*Affiliation:* From Northwest ZooPath, 654 West Main Street, Monroe, WA 98272, USA (Garner M.), San Francisco Zoological Society, 1 Zoo Road, San Francisco, CA 94132, USA (Mutlow), Oakland Zoo, 9777 Golf Links Road, Oakland, CA 94605, USA (Emanualson); Santa Barbara Zoo, 500 Niños Dr., Santa Barbara, CA 93103, USA (Barnes, Garner K.), University of California, Santa Barbara, Animal Resource Center, Santa Barbara, CA 93106, USA (Briggs, Garcia), Fresno Chaffee Zoo, 894 West Belmont Avenue, Fresno, CA 93728, USA (Nodolf).

The California red-legged frog *Rana draytonii,* and yellow-legged frogs *Rana sierrae* and *Rana muscosa*, are federally threatened and endangered species, respectively.[1,2] These frogs are threatened by predation from introduced predators (ie, trout, crayfish and bullfrogs), disease, pesticides, environmental changes due to drought, fire and climate change, and habitat degradation due to livestock grazing, urbanization and connectivity issues (ie, highways). Currently or historically, the San Francisco (SFZ), Fresno-Chaffee (FCZ) and Oakland zoo (OZ), and the University of California, Santa Barbara (UCSB) have been involved in programs to restore wild populations to the Sierra Nevada mountain range of California. These programs have included wild egg collection for captive rearing, attempts at captive propagation, and research on the pathogenesis and treatment of chytrid infection. A review of mortality data for frogs submitted from these facilities was performed from the Northwest ZooPath archives. Total submissions (168, 135 died, 33 euthanized) included yellow-legged frogs (155) and red-legged frogs (13). Total submissions from each institution included SFZ (95), OZ (51), FCZ (12), and UCSB (10). Ages ranged from 1 week to 8 years, including 20 tadpoles and 52 adults of unstated age. Gender included 60 male, 73 female, 1 hermaphrodite and 34 undetermined. The most important disease categories were inflammatory/infectious disease (57%), trauma (28%), degenerative disease (27%) and cardiovascular disease (17%). Bacterial infections (31) included sepsis (9), bronchopneumonia (7), pustular dermatitis (5), coelomitis (4), hepatitis (2) and enteritis (2), and mycobacterial granulomatosis (3). Mycotic infections (22) included chytridiomycosis (8) epidermal saprolegniosis and similar molds (7), and phaeohyphomycosis (5). Parasitism (8) included branchial protozoa (4), encysted trematodes (2), renal myxozoa (1) and pulmonary flagellates (1). Nutritional diseases (30) included inanition (17), nutritional bone disease (6) and squamous metaplasia (primarily renal) (6). The most frequent presentation, encompassing multiple organ system disorders, was renal tubular necrosis (29), associated with generalized edema (26), and various forms of epidermal disease (22).

## References


1. Richmond JQ, Barr KR, Backlin AR, et al. Evolutionary dynamics of a rapidly receding southern range boundary in the threatened California red-legged frog (*Rana draytonii*). *Evol Appl.* 2013;6:808-822.

2. Knapp RA, Fellers GM, Kleeman PM, et al. Large-scale recovery of an endangered amphibian despite ongoing exposure to multiple stressors. *Proc Natl Acad Sci USA.* 2016;113:11889-11894.

# Agreement Among Three Methods to Determine Blood Testosterone Concentrations in Two *Cyclura* species

Sean M. Perry, DVM,
Jon Fletcher, DVM, Dipl ACVIM (Internal Medicine),
Mark A. Mitchell, DVM, MS, PhD, DipECZM (Herpetology)

**Session #2789**

*Affiliation:* From the Department of Veterinary Clinical Sciences, Louisiana State University, School of Veterinary Medicine, Skip Bertman Drive, Baton Rouge, LA 70803, USA.

*Note:* Abstract format selected by author(s). See Front Matter for description of conference proceedings manuscripts submitted as scientific abstracts.

*Abstract:* Determining accurate and precise hormone concentrations in reptiles is difficult due to the numerous species and the need to validate each hormone assay for each respective species. The objective of this study was to determine the level of agreement among three immunoassays to evaluate circulating testosterone concentrations of Grand Cayman rock iguana hybrids *(Cyclura lewisi x nubila)* and rhinoceros rock iguanas (*Cyclura cornuta*). Blood was taken from 23 *Cyclura lewisi x nubila* and *Cyclura cornuta* once at the beginning and end of the reproductive season. A testosterone enzyme immunoassay (EIA) from Arbor Assays (Ann Arbor, MI 48108, USA) was developed for both species to determine circulating testosterone concentrations and is considered the gold standard for evaluating blood testosterone concentrations. Plasma samples were extracted using an ethyl acetate plasma extraction with liquid nitrogen. Parallelism was performed to determine the ideal dilution for analysis. Recovery and accuracy checks were performed with spiked pooled samples. Samples split and were additionally analyzed using 2 commercial high output laboratory analyzers, the Siemens Immulite 1000 (Siemens AG, Muenchen, Germany) and the TOSOH AIA 360 Automated Immunoanalyzer (Tosoh Bioscience, Inc, San Francisco, CA 94080, USA). Agreement analysis was performed using Bland-Altman plots and Passing-Bablok regression to evaluate the level of agreement among the three methodologies.

# A Standardized Approach to Developing Cryopreservation Protocols for Reptile Semen Using the Red-Eared Slider (*Trachemys scripta elegans*) as a Model

Sean M. Perry, DVM, and
Mark A. Mitchell, DVM, MS, PhD, DipECZM (Herpetology)

**Session #2816**

*Affiliation:* From the Department of Veterinary Clinical Sciences, Louisiana State University, School of Veterinary Medicine, Skip Bertman Drive, Baton Rouge, LA 70803, USA.

*Note:* Abstract format selected by author(s). See Front Matter for description of conference proceedings manuscripts submitted as scientific abstracts.

*Abstract:* Germplasm cryopreservation and biobanking in reptiles are lagging behind all other taxa despite reptiles being the third most diverse taxa of vertebrates. Reptiles are unique and diverse in their reproductive strategies and much of our understanding of reptile reproduction remains unknown. Even antemortem semen handling and collection in reptiles have only been recently repeatedly performed. The purpose of this study was to develop a consistent method for developing cryopreservation protocols for reptiles. A total of 18 red-eared sliders (*Trachemys scripta elegans*) were used in three experiments to develop a cryopreservation protocol. First, we determined the best extender for semen in *T. scripta elegans* during cooled storage at 4°C. Six semen extenders (INRA 96, Test Yolk Buffer, F-10, Sperm Wash Media, Electrolyte Free Media, Hanks balanced salt solution) were evaluated for 96 hours. INRA was determined to be the best extender. Second, we evaluated cryoprotectant toxicity with three cryoprotectants (DMSO, glycerol, and methanol) at 5%, 10%, and 15% concentrations with INRA 96 at 20°C over 50 minutes. Based on the cryoprotectant toxicity, DMSO and glycerol were chosen as cryoprotectants at 2%, 4%, and 6% concentrations in INRA 96 for the cryopreservation study. Samples were frozen at three different freezing rates of -10°C, -20°C, -30°C. All semen samples were evaluated with a computer-assisted sperm analyzer and frozen samples also were evaluated with live/dead fluorescence and flow cytometry.

# Investigating the Ability to Detect *Ophidiomyces ophiodiicola* in Snake Sheds

Bryce Dibadj, BA,
Ellen Haynes, DVM,
Kathryn Vivirito, MS,
Allison Wright, MCM,
Kristin Stanford, PhD,
Matthew C. Allender, DVM, MS, PhD, Dipl ACZM

**Session #2800**

*Affiliation:* From the Wildlife Epidemiology Laboratory, University of Illinois College of Veterinary Medicine, 2001 S Lincoln Ave, Urbana, IL 61802, USA (Dibadj, Haynes, Vivirito, Wright, Allender), The Ohio State University Stone Laboratory, 878 Bayview Ave, Put-In-Bay, OH 43456, USA (Stanford).

Ophidiomycosis is an emerging infectious disease caused by the fungus *Ophidiomyces ophiodiicola* that has impacted free-ranging snake populations throughout the United States and Europe. Diagnostic limitations have significantly impeded characterization of the epidemiology of this disease and, subsequently, have impacted efforts to improve clinical care and conservation outcomes. The purpose of this research was to investigate the detection of *O. ophiodiicola* in shed skins from wild Lake Erie watersnakes (*Nerodia sipedon insularum*) with known ophidiomycosis status. We examined 33 shed skin samples from 17 individuals and each shed was evaluated for evidence corresponding to confirmed ophidiomycosis lesions. If evidence was present, a 1 × 2 cm piece of shed including the lesion was removed; if evidence of a lesion was not identified, a random sample of skin measuring the same dimensions was collected. For each sample, DNA was extracted and analyzed using a qPCR assay specific for *O. ophiodiicola*. Results were then compared to qPCR results from a body swab or corresponding lesion swab from the same individual. All (100%) sheds from individuals with confirmed ophidiomycosis were also positive for *O. ophiodiicola*. The quantity of fungus in the shed (median = 139.7 copies/ng DNA) was significantly greater than for the swab collected before the shed (median = 11.57 copies/ng DNA) ($P = 0.011$), but there was no significant difference between the shed and the swab collected after the shed (median = 454.25 copies/ng DNA) ($P = 0.47$). These results suggest that quantitative PCR of snake sheds may be an effective detection method for the positive presence of *O. ophiodiicola*.

# Hematologic Analysis of Wild Lake Erie Water Snakes and the Effect of Sex, Site, and Ophidiomycosis Status

Brina Gartlan, BS,
Ellen Haynes, DVM,
Kathryn Vivirito, MS,
Allison Wright, MS, MCM,
Kristin Stanford, PhD,
Matthew C. Allender, DVM, MS, PhD, Dipl ACZM

**Session # 2819**

*Affiliation:* From the Wildlife Epidemiology Laboratory, University of Illinois, College of Veterinary Medicine, Champaign-Urbana, IL 61802, USA (Gartlan, Haynes, Vivirito, Wright, Allender), Stone Laboratory, The Ohio State University, 878 Bayview Ave, Put-in-Bay, OH 43456, USA (Stanford).

*Note:* Abstract format selected by author(s). See Front Matter for description of conference proceedings manuscripts submitted as scientific abstracts.

*Abstract:* Habitat loss, human interaction, and infectious diseases all threaten declining reptile populations. Lake Erie water snakes, *Nerodia sipedon insularum* (LEWS), once listed as an endangered species, have begun to recover. Hematologic analysis can provide information about the presence of inflammation and infection and thereby inform conservation efforts, but such analysis has not been performed in *N. sipedon insularum* to date. The objective of this study was to evaluate hematologic parameters in LEWS and identify differences based on sex, location, and ophidiomycosis status. Blood was drawn from 78 wild-caught snakes at 6 sites in OH in 2018. PCV was measured using the hematocrit method, while total WBC was estimated from stained blood films and 100 cell count differentials were performed by Brina Gartlan. Packed cell volume (PCV) was significantly higher in males (mean = 31.88) than females (median = 27), while total white blood cell count (WBC) was significantly higher in females (mean = 19,708.33) than males (median = 15,400). Animals with possible or apparent ophidiomycosis has significantly higher azurophils than individuals with the presence of *Ophidiomyces* alone ($P < 0.01$). This was the first study to investigate hematology in a free-ranging population of LEWS and will serve as a baseline for future investigations that aim to improve their conservation efforts. Prior studies characterizing the hematologic response following *Ophisiomyces ophiodiicola* infection have observed subtle changes that may not have been observed with the small sample size in this study, but future studies will build on these results to improve early detection and clinical monitoring.

# Health Status Assessment of Traumatic Injury in Freshwater Turtles

Alison Paige Harrelson Savo, DVM,
Yaxin Zheng,
Yuting Zheng,
Gregory A. Lewbart, MS, VMD

**Session #2833**

*Affiliation:* From the Turtle Rescue Team at the North Carolina State University College of Veterinary Medicine, 1060 William Moore Drive, Raleigh, NC 27607, USA (Savo, Lewbart), North Carolina State University Statistics Graduate Program, 5109 SAS Hall, 2311 Stinson Drive, Raleigh, NC 27695, USA (Zheng, Zheng).

*Note:* Abstract format selected by author(s). See Front Matter for description of conference proceedings manuscripts submitted as scientific abstracts.

*Abstract*: A group of injured yellow-bellied sliders (*Trachemys scripta)* and river cooters (*Pseudemys concinna)* were evaluated for a variety of health values at presentation to the NC State Turtle Rescue Team and prior to release. An i-STAT Portable Clinical Analyzer and CG8+ cartridges were used to determine venous blood gas and biochemical values, the packed cell volume (PCV) and total protein were evaluated using hematocrit tubes and high speed centrifugation, and a differential WBC percentage was determined manually with Diff-Quick stained blood smear slides. Forty-six turtles were sampled on presentation and 23 of those animals were sampled again prior to release. Blood values were analyzed for significant differences between samples collected at presentation and prior to release, as well as differences between surviving and non-surviving turtles. Five variables were identified as significantly different between presenting and recuperated samples: pH, pCO2, Glu, % heterophils, and % eosinophils. When comparing samples between turtles that survived versus those that did not, two variables were identified as being significant prognostic indicators: lactate and PCV. Identification of these significant variables can aid in determining patient prognosis and triage therapy for injured aquatic turtles.

## References

1. Savo APH, Zheng Y, Zheng Y, Lewbart GA. Health status assessment of traumatic injury freshwater turtles. *PLoS ONE.* 2018;13(8):e0202194.

**Gap in pagination due to formatting issues.**

**Page 660**

# Nitrate Toxicity in Freshwater Turtles

Anne Fowler, BSc(Vet)(Hons) BVSc, MANZCVS (Avian Health, Wildlife Health, Unusual Pet Medicine)

**Session #2700**

*Affiliation:* From Adelaide Bird & Exotics Vet Centre, 129 Richmond Rd, Richmond, SA, Australia.

*Abstract:* Freshwater turtles are common pets in Australia. Owner knowledge of appropriate tank care is poor and often leads to repeated 'new tank' cycling as demonstrated in a survey of our clients. Turtles presented with anemia and skin lesions. Diagnostic investigation involved determine severity of anemia and infection. Vitamin A, E and C levels were assayed. Treatment involved improving tank care, antibiotic administration, recombinant EPO and vitamin administration.

## Introduction

Australian freshwater turtles, the Eastern long-neck turtle and Murray River short neck turtle are both commonly held pets in Australia. Pet turtles are frequently kept in glass aquarium kept inside the house and are presented for vet examination due to anorexia, shell and skin abnormalities. As part of the clinical workup of these patients, blood tests and water quality tests are undertaken. Often despite the age of the turtle, most tanks are suffering from 'new tank' syndrome. The high ammonia and nitrite levels appear to impact the health of the turtles. The result is anemia, reduction in vitamin A and E levels compared to wild and healthy captive turtles and infection with possible sepsis.

## Tank Care Survey

A survey of our clientele over the last 2 years revealed that there are few clients who understand the nitrogen cycle, often with the result of high nitrate levels in the water. Clients are not using water conditioners, whose purpose is to remove the chlorine products from tap water. Chlorine is added to the town water for its antibacterial properties. Failing to remove it results in the chlorine affecting the nitrogen-eating bacteria required by the filter to remove nitrogenous waste produced by the turtle.

Water changes were rarely performed at the correct frequency or amount. The most common change was to remove 50% of the tank water every 1-3 months. This would result in significant change to the filter bacteria and so some tanks are cycling as 'new tanks' for more than half of the time.

## Clinical Presentation and Diagnostic Workup

Freshwater turtles commonly present for anorexia and lethargy. Some present with ulcerative necrosis of the distal extremities. In severe cases, individuals present with anorexia, poor body condition and petechiation of the skin in the pre-femoral fossa. Diagnostic workup begins with water testing of the tank water to determine the pH and levels of ammonia, nitrite and nitrate. Blood is collected from the jugular vein to determine PCV, total solids and glucose using a hand-held glucometer. A blood smear is made and determination of an estimated white blood cell count and differential are performed.[1] In cases where sufficient blood has been collected, a biochemistry panel and determination of vitamins A, E, and C are done.

The sick turtles demonstrate moderate to severe anemia, leucocytosis and hypoglycemia. Comparisons of vitamin levels in wild, healthy captive, and sick captive turtles were made and affected sick turtles have vitamin levels below that of healthy turtles.

## Treatment of Affected Turtles

Hospitalization or outpatient treatment is elected based upon presence of severe anemia (PCV < 15%), severe hypoglycemia (< 1 mmol/L) and presence or absence of petechiation. Systemic antibiotics with ceftazidime (20 mg/kg IM q72h × 30d) are commenced with a recheck scheduled at 30 days to retest altered blood parameters. At monthly visits, the turtle is weighed and examined to check skin lesions and change in body condition. A total antibiotic course of 60-90 days duration is expected based upon sequential estimated total white blood cell counts. An injection of vitamin ADEC (Vetafarm, Wagga Wagga, Australia) at 0.1 ml/150 g is given subcutaneously at the monthly visits as vitamin E has been shown to mitigate the effects of nitrite toxicity.[2] A human recombinant erythropoeitin, darbepoetin alfa, (Arasnep, Amgen, Australia) is given at 10 000 IU/kg SC q7d over 4 weeks in cases of moderate to severe anemia. Vitamin B complex containing B9 is also administered subcutaneously when anemia is demonstrated as this vitamin plays a role in the development of erythrocytes. Ulcerative skin lesions are treated daily with topical application of silver sulfasalazine (Flamazine, Smith & Nephew, Australia) until resolution.

Either in the hospital or at home, the turtle is housed out of the tank water for a period during the day. This removes the turtle from the toxic environment and effectively reduces the load on the filter. Dry-docking involves housing the turtle in a large plastic tub with a wet towel as a substrate. The tub has a heat pad placed under half of the length. This generates a heat at 22-24 °C and 70-80% humidity inside the tub.

## Conclusions

Nitrite poisoning causes methemoglobinemia,[3] impairing the ability of blood cells to exchange hemoglobin with resulting anemia as seen in affected turtles. Vitamin levels of A, E[1,4] and C are reduced in nitrate toxicity, due to impaired production, which was demonstrated in sick turtles. High nitrate can cause gastrointestinal ulceration, which may also result in blood loss. Whether it is repeated nitrite spikes or chronic elevations of nitrate that are causing the clinical signs has not been determined. Nitrate may be converted to nitrite in the gut of by anaerobic bacteria in the gut of animals and so a combination may be present.[5] Nitrate toxicity has been documented in a variety of species from fish, frogs, rats and calves. Poor water quality that results in elevated nitrogenous waste in the water of aquatic vertebrates such as fish and frogs has been proven to result in nitrate toxicity, thus this syndrome could be expected to be present in freshwater turtles. Good water quality must be maintained in the tank to ensure the health of pet freshwater turtles.

## References


1. Fudge AM. *Laboratory Medicine: Avian and Exotic Pets*. St. Louis, MO: Elsevier Saunders;2000.

2. Chow CK, Hong CB. Dietary vitamin E and selenium and toxicity of nitrite and nitrate. *Toxicology.* 2002;180(2):195-207.

3. Camargo JA, Alonso A, Salamanca A. Nitrate toxicity to aquatic animals: a review with new data for freshwater invertebrates. *Chemosphere.* 2005;58(9):1255-1267.

4. Atyabi N, Yasini SP, Jalali SM, Shaygan H. Antioxidant effect of different vitamins on methemoglobin production: an in vitro study. *Vet Res Forum.* 2012;3(2):97.

5. Tiso M, Schechter AN. Nitrate reduction to nitrite, nitric oxide and ammonia by gut bacteria under physiological conditions. *PloS One*. 2015;10(3):e0119712.

# Detection of Nidoviruses, *Mycoplasma*, and *Chlamydia* in Pythons

Rachel E. Marschang,
Katharina Racz,
Elisabeth Müller,
Ekaterina Kolesnik

**Session #2664**

*Affiliation:* From Laboklin GmbH & Co. KG, Steubenstr. 4, 97688 Bad Kissingen, Germany.

*Note:* Abstract format selected by author(s). See Front Matter for description of conference proceedings manuscripts submitted as scientific abstracts.

*Abstract:* Nidoviruses have been shown to be an important cause of respiratory disease in pythons. *Mycoplasma* also have been reported in pythons in association with respiratory disease in a few cases, but their clinical importance in these animals is not well understood. *Chlamydia* have been detected in association with a wide range of clinical signs in multiple reptile species. Between January and March 2018, samples from 170 pythons of various species that were submitted to a commercial laboratory for nidovirus detection by PCR also were screened for *Mycoplasma* and *Chlamydia*. The vast majority of the samples tested were oral swabs. Forty-three (25.3%) of the samples were positive for nidoviruses, 101 (59.4%) were positive for *Mycoplasma*, and 3 (1.8%) were positive for *Chlamydia*. Of the nidovirus positive samples, 33 (76.7%) were also *Mycoplasma* positive, significantly more than in snakes that were nidovirus negative (53.7% *Mycoplasma* positive, $P = 0.007411$). There was no correlation between nidovirus and *Chlamydia* infections. While it remains unclear whether *Mycoplasma* play a role as primary respiratory pathogens in pythons, there is a significant correlation between infection with nidoviruses and with *Mycoplasma*. Further study is necessary to determine the clinical effects of these co-infections.

# Snakes in the Suite: Husbandry and Veterinary Care of the Ball Python (*Python regius*) in Biocontainment Laboratories

Lynn J. Miller, DVM, DACLAM

**Session #2630**

*Affiliation:* From the US Army Medical Research Institute of Infectious Diseases, 1425 Porter St., Fort Detrick, MD 21702, USA.

*Note:* Abstract format selected by author(s). See Front Matter for description of conference proceedings manuscripts submitted as scientific abstracts.

*Abstract*: Veterinary care for snakes in a biocontainment laboratory presents several novel challenges. Snakes cannot be treated as if they were ectothermic small mammals. In our study, captive-bred adult ball pythons (*Python regius*) were sourced from reputable, well-known breeders and were quarantined for 60 days before use in the study. During the first half of the quarantine period, all testing was performed to verify the snakes' specific pathogen-free status and overall health. The second half of the quarantine period was conducted in a mock biocontainment lab, allowing the snakes an acclimation period to the smells of protective suits and chemical sanitizers, plus room air exchange rates and negative air pressure. Rack caging setups were most appropriate because they had the best control of microenvironment humidity and temperature. Sanitation measures used in our lab focused on the use of liquid bleach (500-1000 mg/L free chlorine in the final solution), Microchem Plus (1:10 to 1:20 dilution), or paraformaldehyde gas depending on the material to be sanitized. For animal handling, the use of a glove capable of resisting puncture and slash damage, such as Honeywell's Work Easy #300, provided excellent safety. Clear plexiglass shields were useful in controlling animal movement and preventing bite wounds while allowing unimpeded observation of the snakes. Cardiocentesis was the only reliable option to collect a sufficient blood volume for our needs, which was performed on lightly anesthetized animals to facilitate palpation of the heart. In conclusion, safe and effective veterinary care of snake species is possible in high containment laboratories.

# Snapping Turtle Tips, Tricks, and Insights

Renée Schott, DVM, CWR

**Session #2809**

*Affiliation:* From The Wildlife Rehabilitation Center of Minnesota, 2530 Dale St. N, Roseville, MN 55113, USA.

*Note:* Abstract format selected by author(s). See Front Matter for description of conference proceedings manuscripts submitted as scientific abstracts.

*Abstract*: Common snapping turtles (*Chelydra serpentina*) can be intimidating to work with! This lecture presents some tips, tricks and insights that will help make your next common snapping turtle examinations and treatments as easy as working with a slider!

# Biotinylation of Reptilian Erythrocytes

Anke C. Stöhr, DVM, and
Javier Nevarez, DVM, Dipl ACZM, DipECZM (Herpetology)

**Session #2886**

*Affiliation:* From the Department of Veterinary Clinical Sciences, Louisiana State University, Skip Bertman Drive, Baton Rouge, LA 70803, USA.

*Note:* Abstract format selected by author(s). See Front Matter for description of conference proceedings manuscripts submitted as scientific abstracts.

*Abstract:* Blood transfusions can be a valuable treatment option in reptiles, but only limited research has been done to assess its application. One poorly understood aspect is the survival time of reptilian red blood cells in different species. Historically, radioisotope labeling with $^{51}Cr$ or fluorescent labeling has been used to evaluate red blood cells (RBC) survival time in mammalian and nucleated RBCs. Biotin labeling of RBCs, which has been established more recently in mammals to study erythrocyte and transfusion biology, has several advantages: no radiation risks, better precision (no loss of label), better accuracy (no effect of hemodilution/hemoconcentration), capability of long-term labeling, ability to follow multiple RBCs concurrently in one individual, and small sample volume needed. Blood from American alligators was used as a model to establish the biotinylation technique of reptilian RBCs in vitro. Blood was collected in CPDA-1. After washing the RBCs and removal of the buffy coat, the RBC concentration was determined with an automated cell counter and RBCs were labeled with four different concentrations of biotin. Four increasing concentrations of the secondary label (streptavidin-PE) were then added to each group and the RBCs were evaluated via flow cytometry. Two distinct RBC populations could be distinguished based on the density of biotinylation. This is the first description of labeling of nucleated RBCs with biotin. Further studies are planned to confirm the use of this technique in reptiles in vivo and to assess the effect of blood storage and of allogenic blood transfusions on reptilian RBC survival time.

# Investigating the Analytic Sensitivity of Hematologic Methods in Eastern Box Turtles (*Terrapene carolina carolina*)

John Winter

**Session #2787**

*Affiliation:* From University of Illinois College of Veterinary Medicine, 3505 Veterinary Medicine Basic Sciences Building, 2001 S Lincoln Ave, Urbana, IL 61802, USA.

*Note:* Abstract format selected by author(s). See Front Matter for description of conference proceedings manuscripts submitted as scientific abstracts.

*Abstract:* When monitoring the health of reptilian individuals and populations, complete blood counts (CBC) can provide crucial information about inflammation, infection, and stress through the measurement of leukocytes. While multiple methods of measuring leukocytes in reptiles exist, little research has directly compared these methods. In this study, two methods for quantifying reptilian leukocytes, the Avian Leukopet™ kit with hemacytometer (LO) and white blood cell estimates from blood smears (EST), were compared using blood samples from 120 free-ranging eastern box turtles (*Terrapene carolina carolina*) found in Illinois. This study also calculated the analytic variability of the LO method for the first time in reptiles by counting hemacytometers for each sample in triplicate. Significant differences between methods were found with the LO method yielding significantly higher leukocyte numbers than the EST method. The LO method demonstrated an analytic variability of 8.2% inherent to the method itself, independent of biologic variation. A third component of this study was to create a model of biologic variation of the LO and EST method for which each method yielded similar, but not identical results. Leukocytes were significantly affected by age using both LO and EST methods, but leukocyte differences observed between locations was observed only using the EST method. Findings in this study should serve to emphasize the importance of considering hematologic methodology when analyzing CBC results from individuals and populations. The variability in current methods create uncertainty in results and establishment of a gold standard for reptilian leukocyte quantification is needed.

# Evaluation of Two Treatments for *Cryptosporidium serpentis* in Asymptomatic Eastern Indigo Snakes (*Drymarchon couperi*)

James E. Bogan, Jr., DVM, DABVP (Canine and Feline Practice), DABVP (Reptile and Amphibian Practice), CertAqV

**Session #2920**

*Affiliation:* From Central Florida Zoo & Botanical Gardens, 3755 W Seminole Blvd, Sanford, FL 32771, USA.

*Note:* Abstract format selected by author(s). See Front Matter for description of conference proceedings manuscripts submitted as scientific abstracts.

*Abstract:* Thirty-four asymptomatic eastern indigo snakes (*Drymarchon couperi*) naturally infected with *Cryptosporidium serpentis* were randomly divided into two groups (A and B). For Phase 1 of the trial, Group A received 360 mg/kg paromomycin twice weekly in a food item for six weeks and Group B received just the food item. Cloacal PCR swabs were collected every two months for six months to evaluate parasite shedding. PCR negative snakes from both groups were then immunosuppressed with a single dose of 4 mg/kg dexamethasone sodium-phosphate SQ, then PCR screened as described above. PCR negative snakes (six swabs one year from end of treatment) were then re-evaluated via gastric biopsy and PCR analyses. For Group A, 8/17 were cloacal PCR negative prior to immunosuppression compared to 1/17 controls. However, 2/17 Group A snakes were negative (cloacal PCR, gastric PCR, and gastric histology) six months after immunosuppression compared to 1/17 controls. For Phase 2 of the trial, consistently positive snakes from both groups in Phase 1 that had not received dexamethasone SP, were then randomly divided into Groups C and D. Group C received 20 mg/kg nitazoxanide, 10 mg/kg azithromycin, and 5 mg/kg rifabutin twice weekly in a food item for six weeks and Group D received just the food item. Treatment success was evaluated and challenged as with Groups A and B. At the time of the abstract writing, 2/12 snakes from both Groups C and D were cloacal PCR negative for six months. Final results will be reviewed during the presentation.

# Poison Dart Frogs (*Dendrobates tinctorious*) and Leopard Frogs (*Lithobates pipiens*) Exhibit Stress Response to Restraint

Connor Jacobson, Taylor Yaw, DVM, CertAqV,
Christoph Mans, Dr med vet, Dipl ACZM, DipECZM (Zoo Health Management),
Grayson A. Doss, DVM, Dipl ACZM

**Session #2767**

*Affiliation:* From the Department of Surgical Sciences, University of Wisconsin-Madison, School of Veterinary Medicine, Madison, WI 53706, USA.

*Note:* Abstract format selected by author(s). See Front Matter for description of conference proceedings manuscripts submitted as scientific abstracts.

*Abstract*: Frogs are popular exotic pets and often presented for medical care. Manual restraint is commonly used for examining pet amphibians but there is little information on its effects in this group of animals. It has been suggested that amphibians lack emotional responses and therefore consciousness because frogs did not display a stress tachycardia when gently handled as opposed to other animal groups.[1,2] The authors have experienced adverse effects following restraint of debilitated amphibians, however, suggesting that restraint is stressful. To examine if manual restraint results in stress in amphibians, 10 poison dart frogs (*Dendrobates tinctorious*) and 10 leopard frogs (*Lithobates pipiens*) were manually restrained and heart rate (HR) recorded at both the pectoral girdle and pelvic patch using a Doppler probe. Frogs were then randomly placed in 1 of 2 (opaque versus clear) small, plastic containers for 22 minutes and HR measured every 2 minutes through the bottom of the container using the frogs' pelvic patch area. After 22 minutes, frogs were removed from the container, manually restrained again and HR recorded. Heart rate decreased significantly after the first restraint period while frogs were undisturbed in the containers and increased significantly for the second restraint period. There was no significant difference in measured HR between anatomic sites (pelvic versus pectoral) and no clinically relevant difference in HR between container types. Contrary to previous opinion, these results demonstrate that like reptiles, frogs demonstrate a stress tachycardia during manual restraint, and clinicians should be mindful of induced stress when restraining amphibians.

## References

1. Cabanac A, Cabanac M. Heart rate response to gentle handling of frog and lizard. *Behav Processes*. 2000;52:89-95.

2. Cabanac M, Cabanac AJ, Parent A. The emergence of consciousness in phylogeny. *Behav Brain Res*. 2009;198:267-272.

# Evaluating the Efficacy of Alfaxalone in Corn Snakes (*Pantherophis guttatus*)

Kelly Rockwell, DVM,
Kim Boykin, DVM, MS,
Jordan Padlo, CVT,
Courtney Ford,
Mark Mitchell, DVM, MS, PhD, DipECZM (Herpetology)

**Session #2830**

*Affiliation:* From Louisiana State University, School of Veterinary Medicine, 1 Skip Bertman Drive, Baton Rouge, LA 70803, USA.

*Note:* Abstract format selected by author(s). See Front Matter for description of conference proceedings manuscripts submitted as scientific abstracts.

*Abstract*: Alfaxalone is a popular veterinary anesthetic; however, research on the effects of this injectable anesthetic in snakes has been limited to ball pythons (*Python regius*), garter snakes (*Thamnophis sirtalis*), and several Australian snake species. The purpose of this study was to evaluate the anesthetic effects of alfaxalone in corn snakes (*Pantherophis guttatus*), a popular pet snake species. A randomized prospective crossover study was performed with a one-week washout period in between doses. Snakes were randomly assigned to treatment groups and were administered one of three different doses of alfaxalone subcutaneously in the cranial third of the body: 5 mg $kg^{-1}$, 10 mg $kg^{-1}$, and 15 mg $kg^{-1}$. Heart rate, respiratory rate, righting reflex, escape response, tail pinch, needle prick, and tongue flick were monitored at baseline and five-minute intervals until the animal had fully recovered. There was a significant difference in duration of anesthesia, with higher doses producing a longer duration of anesthesia than lower doses. All snakes, except two, lost all reflexes at the 15 mg $kg^{-1}$ dose, and all snakes maintained tail pinch for at least one time point. There was no significant change in heart rate but there was a significant change in respiratory rate based on time but not on dose. No snakes became apneic during the trials. Based on these results, alfaxalone at 15 mg $kg^{-1}$ provides adequate sedation and anesthesia for brief procedures or intubation; however, addition of analgesia is required for painful procedures.

# Specific Anatomy and Contrast Radiography of the Gastrointestinal Tract in Inland Bearded Dragons (*Pogona vitticeps*)

K. Mathes, DVM, Dr med vet, Cert Spec Reptiles, DipECZM (Herpetology),
E. Engelke, DVM, Dr med vet,
Katharina Radelof, DVM, Dr med vet,
C. Pfarrer, DVM, Dr med vet, Prof,
M. Fehr, DVM, Dr med vet, Prof, DipECZM (Small Mammals)

**Session #2724**

*Affiliation:* From Clinic for Small Mammals, Reptiles and Birds, University of Veterinary Medicine Hannover, Foundation, Hannover, Buenteweg 9, Germany (Mathes, Radelof, Fehr), Institute of Anatomy, University of Veterinary Medicine Hannover, Foundation, Hannover, Germany (Engelke, Pfarrer).

*Note:* Abstract format selected by author(s). See Front Matter for description of conference proceedings manuscripts submitted as scientific abstracts.

*Abstract:* In this study, 11 bearded dragon (*Pogona vitticeps*) cadavers (6 females, 5 males) were dissected and examined. Measurements, and documentation of the specific gross anatomy of the alimentary canal were conducted. Topography and location within the coelom of the different parts of the digestive tract including the oropharynx, esophagus, stomach, small intestine (duodenum, jejunum, and ileum), ampulla coli, isthmus coli, rectum, and the distance from the coprodeum to the vent opening were documented. Measurements included snout-vent length, total length of the alimentary canal, and the lengths of the individual sections of the gastrointestinal tract. The ratios of the digestive tract to the snout-vent length (RSVL) were calculated. The distributions of the data were evaluated using the Shapiro-Wilk test. Findings aided in the interpretation of subsequent contrast radiography performed in 22 bearded dragons (9 males, 13 females). Contrast passage was conducted with 2 different contrast media, barium sulfate ($BaSO_4$, Barilux suspension) and an iodinated ionic radiocontrast agent (sodium meglumine amidotrizoate [SMAT], Gastrografin). Radiographic images were obtained from each dragon in both dorsoventral and lateral projections using a digital system whereby all bearded dragons were positioned in ventral recumbency. Knowledge achieved with the dissections helped to accurately identify segments of the lizards' digestive tract filled with contrast medium and an exact transit time could be obtained for the individual parts of the gastrointestinal tract illustrated in the contrast agent passage.

# Normal Ocular Parameters in Healthy Adult Bearded Dragons (*Pogona vitticeps*)

Elyse M. Salpeter, DVM,
Krista A. Keller, DVM, Dipl ACZM,
Bianca C. Martins, DVM, MS, PhD, DACVO,
Kathryn M. Fleming, DVM, PhD, DACVO

**Session #2776**

*Affiliation:* From the College of Veterinary Medicine, University of Illinois, Champaign-Urbana, IL 61802, USA.

The purpose of this study was to describe normal ocular parameters in healthy adult bearded dragons (*Pogona vitticeps*). Ten healthy adult bearded dragons received a full ophthalmic examination including slit-lamp biomicroscopy and indirect fundoscopy. Palpebral lengths were measured using Jameson calipers, tear production was assessed with paper point tear tests (PPTT), intraocular pressures were measured at a single time point with rebound tonometry, corneal touch threshold (CTT) was measured using Cochet-Bonnet aesthesiometery, and central corneal thickness (CCT) was measured using corneal pachymetry. On ophthalmic examination, nictitation involved movement of both upper and lower palpebrae, with more involvement of the lower lid. Palpebral margins were lined by small spicules, and a nictitating membrane was constantly present. Anterior segment findings were consistent, including a circular pupil and iridal vasculature forming an inverted "U" shape around the pupil. Posterior segment findings consistently included an anangiotic retina, absent tapetum and a dark conus papillaris. Mean palpebral length was 6.35 ± 0.81 mm OD (5-7 mm), and 6.25 ± 0.72 mm OS (5-7.5 mm). Mean PPTT was 5.4 ± 1.4 mm/30s OD (4-8 mm/30s) and 4.5 ± 1.6 mm/30s OS (3-8 mm/30s). Mean IOP was 4.9 ± 2.3 mmHg OD (2-9 mmHg) and 5.1 ± 2.3 mmHg OS (2-9 mmHg). Mean CCT was 251.4 ± 9.6 µm OD (242-273 µm) and 249.5 ± 6.4 µm OS (236-258 µm). Mean CTT was 0.9 ± 0.35 g/mm$^2$ OD (0.5-1.8 g/mm$^2$) and 0.9 ± 0.35 g/mm$^2$ (0.5-1.8 g/mm$^2$). In conclusion, this study establishes the normal ocular exam findings and parameters for healthy adult bearded dragons and findings may be of reference when comparing to diseased individuals.

# AUTHOR INDEX